Introduction to Wind Energy

Introduction to Wind Energy

Jayce McCarthy

Callisto Reference

www.callistoreference.com

Callisto Reference,
118-35 Queens Blvd., Suite 400,
Forest Hills, NY 11375, USA

Visit us on the World Wide Web at:
www.callistoreference.com

ISBN: 978-1-64116-591-4 (Hardback)

Cataloging-in-Publication Data

Introduction to wind energy / Jayce McCarthy.
p. cm.
Includes bibliographical references and index.
ISBN 978-1-64116-591-4
1. Wind power. 2. Renewable energy sources. I. McCarthy, Jayce.
TJ820 .I58 2021
333.92--dc23

Table of Contents

Preface

The usage of air flow to rotate wind turbines and thereby turn electric generators is known as wind energy. It is a variable source of energy since there is considerable fluctuation in the amount of power generated over short time scales. Thus, it is used in conjunction with other electric power sources in order to provide steady electrical supply. It can also be connected to batteries. Wind energy is a renewable source of power, which has an extremely low global warming potential per unit of electrical energy generated. Wind energy converter, or wind turbine is a device which converts wind's kinetic energy into mechanical energy. This book is a valuable compilation of topics, ranging from the basic to the most complex advancements in the field of wind energy. It is appropriate for students seeking detailed information in this area as well as for experts.

A foreword of all Chapters of the book is provided below:

Chapter 1 - The natural movement of air in the form of an air current on the surface of Earth is known as wind. The kinetic energy of the wind is called as wind energy by which wind generates electricity. This is an introductory chapter which will briefly introduce all the significant areas related to wind and wind energy; **Chapter 2** - Wind turbines are the devices used for the conversion of kinetic energy of wind into electrical energy. The two main types of wind turbines are horizontal-axis turbines and vertical-axis turbines. The topics elaborated in this chapter will help in gaining a better perspective about these types of wind turbines; **Chapter 3** - Wind turbine design includes rotor, direct current generator or alternating current generator which are used to define the specifications of a wind turbine for extracting wind energy. Aerodynamic modeling, classic barrel design of Savonius, theory of wind turbine and Betz coefficient, etc. are some of the concepts that fall under its domain. This chapter closely examines these concepts of wind turbine to provide an extensive understanding of the subject; **Chapter 4** - Wind energy has a wide range of applications in wind hybrid systems, wind water pumping systems, electricity generation, wind battery charges, wind electricity generators, etc. This chapter has been carefully written to provide an easy understanding of these applications of wind energy; **Chapter 5** - Wind power can have adverse impacts on humans, wildlife and the environment. It includes degradation of the habitat of wildlife, plants and fish, threat to flying animals, noise pollution, visual impacts on landscape, etc. This chapter has been carefully written to provide an easy understanding of these impacts of wind power.

I would like to thank the entire editorial team who made sincere efforts for this book and my family who supported me in my efforts of working on this book. I take this opportunity to thank all those who have been a guiding force throughout my life.

Jayce McCarthy

Chapter 1

Wind Energy: An Introduction

The natural movement of air in the form of an air current on the surface of Earth is known as wind. The kinetic energy of the wind is called as wind energy by which wind generates electricity. This is an introductory chapter which will briefly introduce all the significant areas related to wind and wind energy.

Wind energy (or wind power) describes the process by which wind is used to generate electricity. Wind turbines convert the kinetic energy in the wind into mechanical power. A generator can convert mechanical power into electricity. Mechanical power can also be utilized directly for specific tasks such as pumping water.

Wind is caused by the uneven heating of the atmosphere by the sun, variations in the earth's surface, and rotation of the earth. Mountains, bodies of water, and vegetation all influence wind flow patterns. Wind turbines convert the energy in wind to electricity by rotating propeller-like blades around a rotor. The rotor turns the drive shaft, which turns an electric generator. Three key factors affect the amount of energy a turbine can harness from the wind: wind speed, air density, and swept area.

Equation for Wind Power

$$P = \frac{1}{2}\rho A V^3$$

Wind Speed

The amount of energy in the wind varies with the cube of the wind speed, in other words, if the wind speed doubles, there is eight times more energy in the wind $\left(2^3 = 2x2x2 = 8\right)$. Small changes in wind speed have a large impact on the amount of power available in the wind.

Density of the Air

The more dense the air, the more energy received by the turbine. Air density varies with elevation and temperature. Air is less dense at higher elevations than at sea level, and warm air is less dense than cold air. All else being equal, turbines will produce more power at lower elevations and in locations with cooler average temperatures.

Swept Area of the Turbine

The larger the swept area (the size of the area through which the rotor spins), the more power the turbine can capture from the wind. Since, swept area is $A = pir^2$, where r = radius of the rotor, a small increase in blade length results in a larger increase in the power available to the turbine.

Wind Farm Development

Siting a wind farm varies from one location to another, but there are some important matters for land owners to consider:

- Understand your wind resource.
- Evaluate distance from existing transmission lines.
- Determine benefits of and barriers to allowing your land to be developed.
- Establish access to capital.
- Identify reliable power purchaser or market.
- Address siting and project feasibility considerations.
- Understand wind energy's economics.
- Obtain zoning and permitting expertise.
- Establish dialogue with turbine manufacturers and project developers.
- Secure agreement to meet O&M needs.

The wind is a source of free energy which has been used since ancient times in windmills for pumping water or grinding flour. The technology of high power, geared transmissions was developed centuries ago by windmill designers and the fantail wheel for keeping the main sales pointing into the wind was one of the world's first examples of an automatic control system.

Windmills at Kinderdijk in the Netherlands.
Dating from 1740, used for Pumping Water from the Polder.

Though modern technology has made dramatic improvements to the efficiency of windmills which are now extensively use for electricity generation, they are still dependent on the vagaries of the weather. Not just on the wind direction but on the intermittent and unpredictable force of the wind. Too little wind and they can't deliver sufficient sustained power to overcome frictional losses in the system. Too much and they are susceptible to damage. Between these extremes, cost efficient installations have been developed to extract energy from the wind.

Available Power From the Wind

Theoretical Power

The power P available in the wind impinging on a wind driven generator is given by:

$$P = \frac{1}{2}CA\rho v^3$$

where, C is an efficiency factor known as the Power Coefficient which depends on the machine design, A is the area of the wind front intercepted by the rotor blades (the swept area), ρ is the density of the air (averaging 1.225 Kg/m^3 at sea level) and v is the wind velocity.

Note that the power is proportional to area swept by the blades, the density of the air and to the cube of the wind speed. Thus doubling the blade length will produce four times the power and doubling the wind speed will produce eight times the power.

Note also that the effective swept area of the blades is an annular ring, not a circle, because of the dead space around the hub of the blades. A similar equation applies to the theoretical power generated by a "run of river" and "tidal flow" hydro turbines.

Energy Conversion

Practical Power and Conversion Efficiency

German aerodynamicist Albert Betz showed that a maximum of only 59.3% of the theoretical power can be extracted from the wind, no matter how good the wind turbine is, otherwise the wind would stop when it hit the blades. He demonstrated mathematically that the optimum occurs when the rotor reduces the wind speed by one third.

In practical designs, inefficiencies in the design and frictional losses will reduce the power available from the wind still further. Converting this wind power into electrical power also incurs losses of up to 10% in the drive train and the generator and another 10% in the inverter and cabling. Furthermore, when the wind speed exceeds the rated wind speed, control systems limit the energy conversion in order to protect the electric generator so that ultimately, the wind turbine will convert only about 30% to 35% of the available wind energy into electrical energy.

Note that the power output from commercially available domestic wind turbines is usually specified at a steady, gust free, wind speed of 12.5 m/s. (Force 6 on the Beaufort scale corresponding to a strong breeze). In many locations, particularly urban installations, the prevailing wind will rarely reach this speed.

Blade Design for Optimum Energy Capture

Modern, high capacity wind turbines, such as those used by the electricity utilities in the electricity grid, typically have blades with a cross section similar to the aerofoils used to provide the lift in aircraft wings.

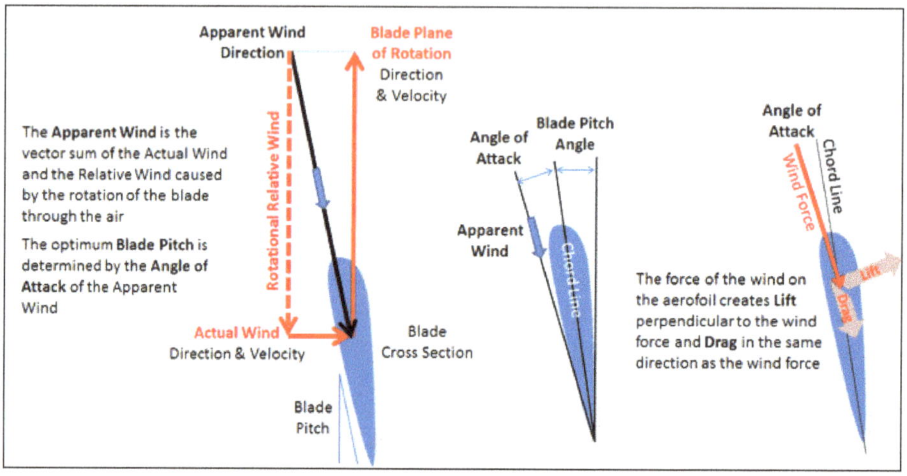

Turbine blade aerodynamics.

The direction of the apparent wind, that is the incident wind, relative to the chord line of the aerofoil is known as the angle of attack. Just as with aircraft wings, the lift resulting from the incident wind force increases as the angle of attack increases from 0 to a maximum of about 15 degrees at which point the smooth laminar flow of the air over the blade ceases and the air flow over the blade separates from the aerofoil and becomes turbulent. Above this point the lift force deteriorates rapidly while drag increases leading to a stall.

The tangential velocity S of any blade section at a distance r from the centre of rotation (the root of the blade) is given by $S = r\Omega$ where Ω is the angular velocity of rotation in radians. For a given wind speed the apparent wind will be different at the root of the blade from the apparent wind at the tip of the blade because the rotational relative wind speed is different.

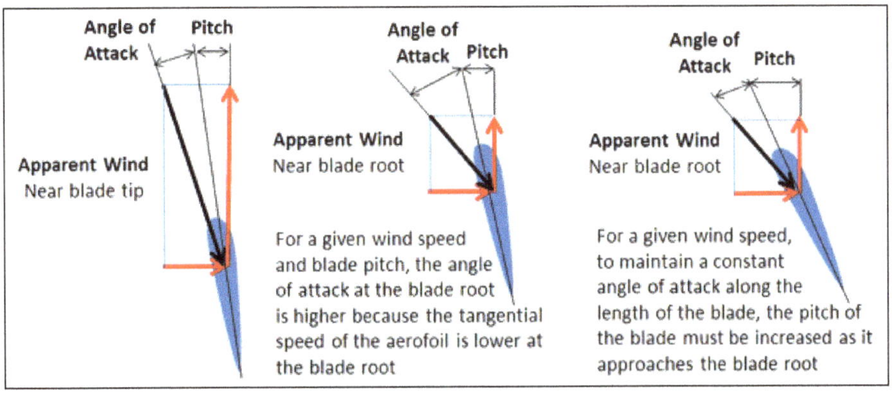

Angle of attack and blade twist.

For a given speed of rotation, the tangential velocity of sections of the blade increases along the length of the blade towards the tip, so that the pitch of the blade must be twisted to maintain the same, optimum angle of attack at all sections along the length of the blade. The blade twist is thus, optimised for a given wind speed. As the wind speed changes however, the twist will no longer be optimum. To retain the optimum angle of attack as wind speed increases a fixed pitch blade must increase its rotational speed accordingly, otherwise, for fixed speed rotors, variable pitch blades must be used.

The number of blades in the turbine rotor and its rotational speed must be optimised to extract the maximum energy from the available wind.

While using rotors with multiple blades should capture more wind energy, there is a practical limit to the number of blades which can be used because each blade of a spinning rotor leaves turbulence in its wake and this reduces the amount of energy which the following blade can extract from the wind. This same turbulence effect also limits the possible rotor speeds because a high speed rotor does not provide enough time for the air flow to settle after the passage of a blade before the next blade comes along.

There is also a lower limit to both the number of blades and the rotor speed. With too few rotor blades, or a slow turning rotor, most of the wind will pass undisturbed through the gap between the blades reducing the potential for capturing the wind energy. The fewer the number of blades, the faster the wind turbine rotor needs to turn to extract maximum power from the wind.

The notion of the Tip Speed Ratio (TSR) is a concept used by wind turbine designers to optimise a blade set to the shaft speed required by a particular electricity generator while extracting the maximum energy from the wind. The tip speed ratio is given by:

$$TSR = \Omega R/V$$

Where, Ω is the angular velocity of the rotor, R is the distance between the axis of rotation and the tip of the blade, and V is the wind speed.

A well designed typical three-bladed rotor would have a tip speed ratio of around 6 to 7.

Design Limits

For safety and efficiency reasons wind turbines are subject to operating limits depending on the wind conditions and the system design.

- Cut - in Wind Speed: This is the minimum wind velocity below which no useful power output can be produced from wind turbine, typically between 3 and 4 m/s (10 and 14 km/h, 7 and 9 mph).
- Rated Wind Speed (also associated with the Nameplate Capacity): This is the lowest wind velocity at which the turbine develops its full power. This corresponds to the maximum, safe electrical generating capacity which the associated electrical generator can handle, in other words the generator's rated electrical power output. The rated wind speed is typically about 15 m/s (54 km/h, 34 mph) which is about double the expected average speed of the wind. To keep the turbine operating with wind speeds above the rated wind speed, control systems may be used to vary the pitch of the turbine blades, reducing the rotation speed of the rotor and thus limiting the mechanical power applied to the generator so that the electrical output remains constant. Though the turbine works with winds speeds right up to the cut-out wind speed, its efficiency is automatically reduced at speeds above the rated speed so that it captures less of the available wind energy in order to protect the generator. While it would be possible to use larger generators to extract full power from the wind at speeds over the rated wind speed, this would not normally be economical because of the lower frequency of occurrence of wind speeds above the rated wind speed.
- Cut-out Wind Speed: This is the maximum safe working wind speed and the speed at which the wind turbine is designed to be shut down by applying brakes to prevent damage to the system. In addition to electrical or mechanical brakes, the turbine may be slowed down by stalling or furling.
    - Stalling: This is a self correcting or passive strategy which can be used with fixed speed wind turbines. As the wind speed increases so does the wind angle of attack until it reaches its stalling angle at which point the "lift" force turning the blade is destroyed. However, increasing the angle of attack also increases the effective cross section of the blade face-on to the wind, and thus the direct

wind force and the associated stress on the blades. A fully stalled turbine blade, when stopped, has the flat side of the blade facing directly into the wind.

- Furling or Feathering: This is a technique derived from sailing in which the pitch control of the blades is used to decrease the angle of attack which in turn reduces the "lift" on the blades as well as the effective cross section of the aerofoil facing into the wind. A fully furled turbine blade, when stopped, has the edge of the blade facing into the wind reducing the wind force and stresses on the blade.

The cut-out speed is specified to be as high possible consistent with safety requirements and practicality in order to capture as much as possible of the available wind energy over the full spectrum of expected wind speeds. A cut-out speed of 25 m/s (90 km/h, 56 mph) is typical for very large turbines.

- Survival Wind Speed: This is the maximum wind speed that a given wind turbine is designed to withstand above which it cannot survive. The survival speed of commercial wind turbines is in the range of 50 m/s (180 km/h, 112 mph) to 72 m/s (259 km/h, 161 mph). The most common survival speed is 60 m/s (216 km/h, 134 mph). The safe survival speed depends on local wind conditions is usually regulated by national safety standards.

Yaw Control

Windmills can only extract the maximum power from the available wind when the plane of rotation of the blades is perpendicular to the direction of the wind. To ensure this the rotor mount must be free to rotate on its vertical axis and the installation must include some form of yaw control to turn the rotor into the wind.

For small, lightweight installations this is normally accomplished by adding a tail fin behind the rotor in line with its axis. Any lateral component of the wind will tend to push the side of the tail fin causing the rotor mount to turn until the fin is in line with the wind. When the rotor is facing into the wind there will be no lateral force on the fin and the rotor will remain in position. Friction and inertia will tend to hold it in position so that it does not follow small disturbances. Large turbine installations have automatic control systems with wind sensors to monitor the direction of the wind and a powered mechanism to drive the rotor into its optimum position.

Capacity Factor

Electrical generating equipment is usually specified at its rated capacity. This is normally the maximum power or energy output which can be generated in optimal conditions. Since a wind turbine rarely works at its optimal capacity the actual energy output over a year will be much less than its rated capacity. Furthermore there will often be periods when the wind turbine cannot deliver any power at all. These occur when there is insufficient wind to power the turbine system, or other periods, fortunately only a few, when the wind turbine must be shut down because the wind speed is dangerously high and exceeds the system cut-out speed.

The capacity factor is simply the wind turbine generator's actual energy output for a given period divided by the theoretical energy output if the machine had operated at its rated power output for the same period. Typical capacity factors for wind turbines range from 0.25 to 0.30. Thus a wind turbine rated at 1 MegaWatt will deliver on average only about 250 kiloWatts of power. (For comparison, the capacity factor of thermal power generation is between 0.70 and 0.90).

Wind Supply Characteristics

Wind Speed

Though the force and power of the wind are difficult to quantify, various scales and descriptions have been used to characterise its intensity. The Beaufort scale is one measure in common use. The lowest point or zero on the Beaufort scale corresponds to the calmest conditions when the wind speed is zero and smoke rises vertically. The highest point is defined as force 12 when the wind speed is greater than 34 metres per second (122 km/h, 76 mph). as occurs in tropical cyclones when the countryside is devastated by hurricane conditions.

Small wind turbines generally operate between force 3 and force 7 on the Beaufort scale with the rated capacity commonly being defined at force 6 with a wind speed of 12 m/s. Below force 3 the wind turbine will not generate significant power.

At force 3, wind speeds range from 3.6 to 5.8 m/s (8 to 13 mph). Wind conditions are described as "light" and leaves are in movement and flags begin to extend.

At force 7, wind speeds range from 14 to 17 m/s (32 to 39 mph). Wind conditions are described as "strong" and whole trees are in motion.

With winds above force 7 small, domestic wind turbines should be shut down to prevent damage.

Large turbines used in the electricity grid are designed to work with wind speeds of up to 25 m/s (90 km/h, 56 mph) which corresponds to between force 9 (severe gale, 23 m/s) and force 10 (storm, 27 m/s) on the Beaufort Scale.

Wind Consistency

Wind power has the advantage that it is normally available 24 hours per day, unlike solar power which is only available during daylight hours. Unfortunately the availability of wind energy is less predictable than solar energy. At least we know that the sun rises and sets every day. Nevertheless, based on data collected over many years, some predictions about the frequency of the wind at various speeds, if not the timing, are possible.

Wind Speed Distribution

Care should be taken in calculating the amount of energy available from the wind as it is quite common to overestimate its potential. You can not simply take the *average* of the wind speeds throughout the year and use it to calculate the energy available from the wind because its speed is constantly changing and its power is proportional to the cube of the wind speed. (Energy = Power X Time). You have to weigh the probability of each wind speed with the corresponding amount of energy it carries.

Experience shows that for a given height above ground, the frequency at which the wind blows with any particular speed follows a Rayleigh Distribution. An example is shown below:

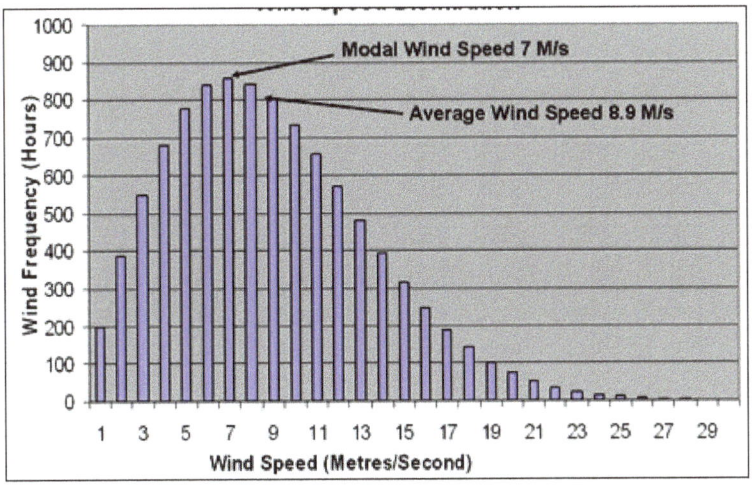

Wind speed distribution.

- The modal wind speed, that is the speed at which the wind most frequently blows, is less than the average wind speed which is the speed often quoted as representing the typical wind conditions. For reference, the average wind speed across the UK quoted by the Department of Trade and Industry (DTI), is approximately 5.6 metres per second [m/s] at 10 metres above ground level (agl)".
- Published average wind speeds are only reliable for open rural environments. Wind speeds just above roof level in urban environments will be considerably less than the quoted averages because of turbulence and shielding caused by buildings and trees. A wind turbine sited below the ridge of a building or at a similar height in the garden of an urban dwelling as often shown in the product sales literature is unlikely to provide the energy levels claimed in the specifications.
- The distribution does not represent the energy content of the wind since this is proportional to the cube of the wind speed.
- A distribution such as the one above is only valid for the prevailing wind conditions at a particular height above the ground. Average wind speeds usually tend

to increase with height then level off which is why wind turbines are usually installed as high above ground as possible.

An empirical formula developed by D.L. Elliott of Pacific Northwest Labs gives the wind speed V at a height H above ground level as,

$$V = Vref\,(H / Href)^{\alpha}$$

Where Vref is the reference wind speed at a reference height Href and the exponent α is a correction factor dependent on obstacles on the ground, the density of the air and wind stability factors. In wind resource assessments α is commonly assumed to be a constant 1/7th. The histogram below shows this relationship.

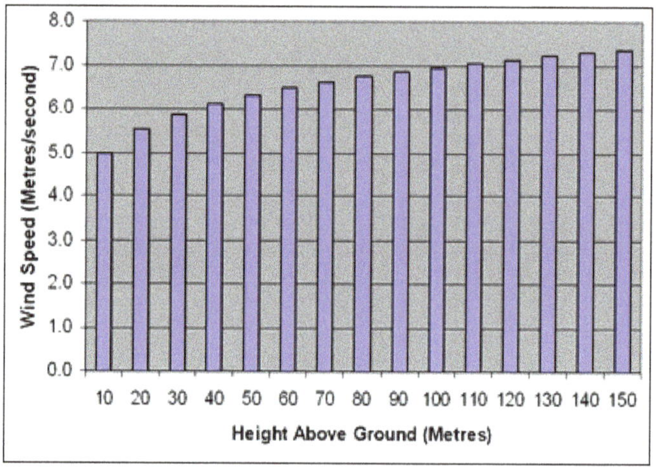

Wind speed vs Height above ground.

Wind Energy Distribution

The histogram below shows the resulting distribution of the wind energy content superimposed on the Rayleigh wind speed distribution (above) which caused it. Unfortunately not all of this wind energy can be captured by conventional wind turbines.

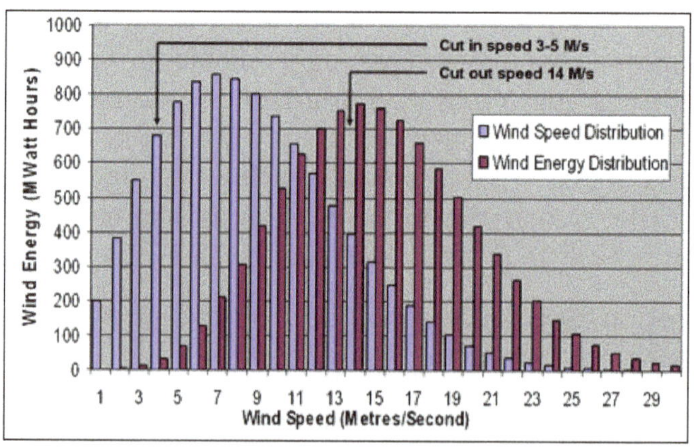

Wind energy content distribution.

- The peak wind energy occurs at wind speeds considerably above both the modal and average wind speeds since the wind energy content is proportional to the cube of its speed.
- Very little energy is available at low speeds and most of this will be needed to overcome frictional losses in the wind turbine. Energy generation typically does not cut in until wind is blowing at speeds of at least 3 m/s to 5 m/s.
- High wind speeds cause high rotation speeds and high stresses in the wind turbine which can can result in serious damage to the installation. To avoid these dangerous conditions, wind turbines are usually designed to cut out at wind speeds of around 25 m/s either by braking or feathering the rotor blades allowing the wind to spill over the blades, though smaller domestic installations may have lower operating limits.
- Because of the limitations of the generating system and also upper speed limit at which the wind turbine can safely be used, it may capture only half or less of the available wind energy.

For a given wind speed the wind energy also depends on the elevation of the wind turbine above sea level. This is because the density of the air decreases with altitude and the wind energy is proportional to the air density. This effect is shown in the following histogram.

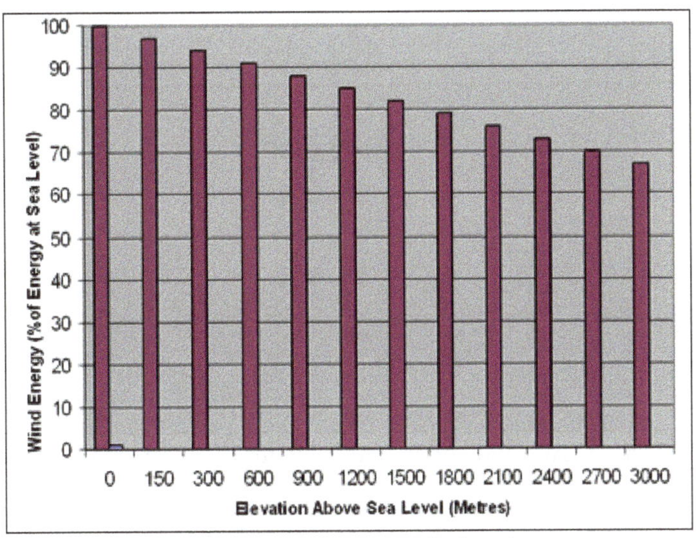

Wind energy reduction with elevation.

- For a given wind speed the wind energy density decreases with increases in altitude. However at the same time the actual wind speeds tend to increase with height above ground level. Since the wind energy is proportional to the cube of the wind speed, the net effect is that wind energy tends to increase with the height above ground level.

- As the density of air decreases with altitude, the wind energy density also decreases. By contrast the available solar energy increases with altitude due to lower atmospheric absorption.

Location Considerations

Generally marine locations and exposed hilltops provide the most favourable wind conditions with wind speeds consistently greater than 5 m/s.

Turbulent conditions will reduce the amount of energy which can be extracted from the wind reducing in turn the overall efficiency of the system. This is more likely to be the case over land than over the sea. Raising the height of the turbine above the ground effectively lifts it above the worst of the turbulence and improves efficiency.

Domestic wind turbines located between buildings in urban environments rarely operate at peak efficiency suffering from turbulence as well as being shielded from the wind by buildings and trees.

Practical Systems

Community/Grid Installations

Vesta 7 MW Wind Turbines with a Rotor Diameters of 164 m.

Grid connected systems are dimensioned for average wind speeds 5.5 m/s on land and 6.5 m/s offshore where wind turbulence is less and wind speeds are higher. While offshore plants benefit from higher sustainable wind speeds, their construction and maintenance costs are higher.

Large scale wind turbine generators with outputs of up to 8 MWe or more with rotor diameters up to164 metres are now functioning in many regions of the world with even larger designs in the pipeline.

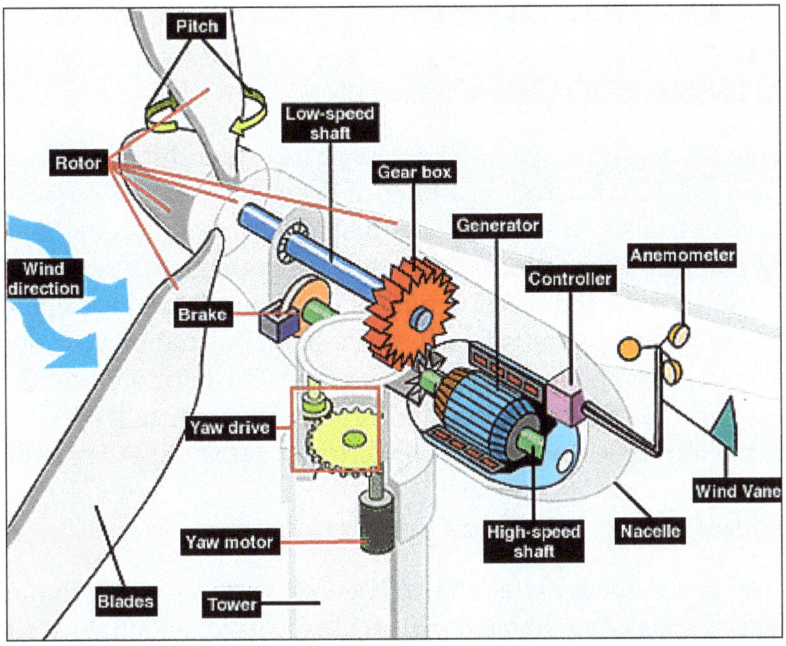

Large rotor blades are necessary to intercept the maximum air stream but these give rise to very high tip speeds. The tip speeds however must be limited, mainly because of unacceptable noise levels, resulting in very low rotation speeds which may be as low as 10 to 20 rpm for large wind turbines. The operating speed of the generator is however is much higher, typically 1200 rpm, determined by the number of its magnetic pole pairs and the frequency of the grid electrical supply. Consequently a gearbox must be used to increase the shaft speed to drive the generator at the fixed synchronous speed corresponding to the grid frequency.

Note that a "synchronous generator" is one whose electrical output frequency is synchronised to its shaft speed. It is not necessarily synchronised to the grid frequency, although that is usually an objective and extra, external controls are necessary to achieve this.

Fixed Speed Wind Turbine Generators

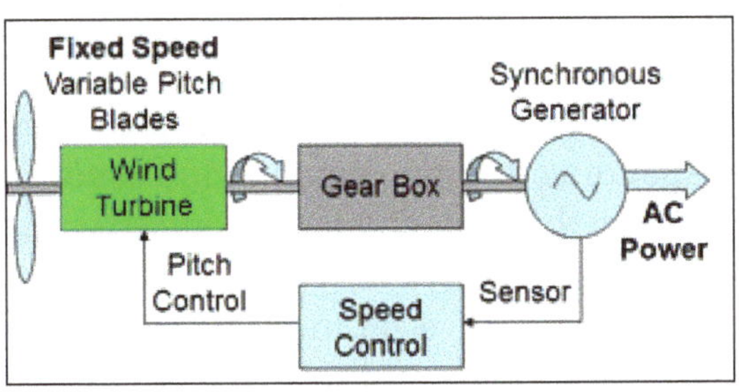

Large scale wind power (Grid systems).

A typical fixed speed system employs a rotor with three variable pitch blades which are controlled automatically to maintain a fixed rotation speed for any wind speed. The rotor drives a synchronous generator through a gear box and the whole assembly is housed in a nacelle on top of a substantial tower with massive foundations requiring hundreds of cubic metres of reinforced concrete.

Fixed speed systems may however suffer excessive mechanical stresses. Because they are required to maintain a fixed speed regardless of the wind speed, there is no "give" in the mechanism to absorb gusty wind forces and this results in high torque, high stresses and excessive wear and tear on the gear box increasing maintenance costs and reducing service life. At the same time, the reaction time of these mechanical systems can be in the range of tens of milliseconds so that each time a burst of wind hits the turbine, a rapid fluctuation of electrical output power can be observed. Furthermore, variable speed wind turbines can capture 8-15% more of the wind's energy than constant speed machines. For these reasons, variable speed systems are preferred over fixed speed systems.

Variable Speed Wind Turbine Generators

A variable speed generator is better able to cope with stormy wind conditions because its rotor can speed up or slow down to absorb the forces when bursts of wind suddenly increase the torque on the system. The electronic control systems will keep the generator's output frequency constant during these fluctuating wind conditions.

Synchronous Generator with In-Line Frequency Control

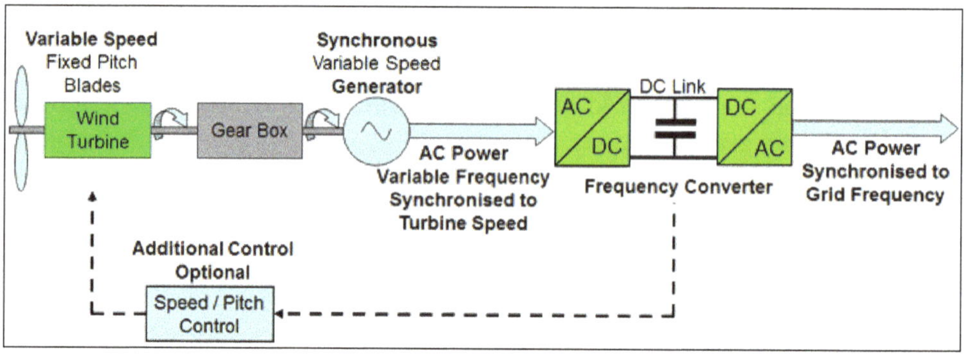

Large scale wind power with In-line frequency conversion (grid systems).

Rather than controlling the turbine rotation speed to obtain a fixed frequency synchronised with the grid from a synchronous generator, the rotor and turbine can be run at a variable speed corresponding to the prevailing wind conditions. This will produce a varying frequency output from the generator synchronised with the drive shaft rotation speed. This output can then be rectified in the generator side of an AC-DC-AC converter and the converted back to AC in an inverter in grid side of the converter which is synchronised with the grid frequency. The grid side converter can also be used to provide reactive power (VArs) to the grid for power factor control and voltage regulation by varying the firing angle of the thyristor switching in the inverter and thus the phase of the output current with respect to the voltage.

The range of wind speeds over which the system can be operated can be extended and mechanical safety controls can be incorporated by means of an optional speed control system based on pitch control of the rotor vanes as used in the fixed speed system.

One major drawback of this system is that the components and the electronic control circuits in the frequency converter must be dimensioned to carry the full generator power. The doubly fed induction generator DFIG overcomes this difficulty.

Doubly Fed Induction Generator - DFIG

DFIG technology is currently the preferred wind power generating technology. The basic grid connected asynchronous induction generator gets its excitation current from the grid through the stator windings and has limited control over its output voltage and frequency. The doubly fed induction generator permits a second excitation current input, through slip rings to a wound rotor permitting greater control over the generator output.

The DFIG system consists of a 3 phase wound rotor generator with its stator windings fed from the grid and its rotor windings fed via a back to back converter system in a bidirectional feedback loop taking power either from the grid to the generator or from the generator to the grid.

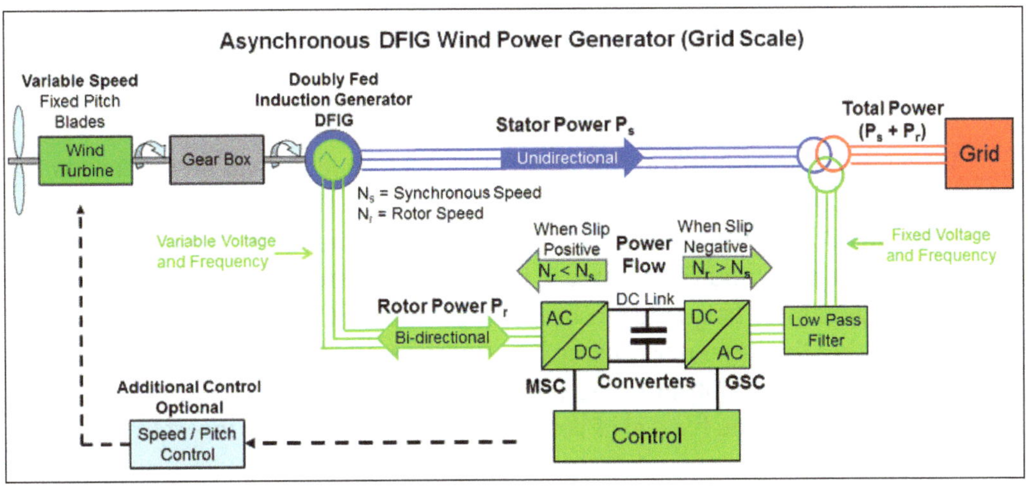

Generator Operating Principle

The feedback control system monitors the stator output voltage and frequency and provides error signals if these are different from the grid standards. The frequency error is equal to the generator slip frequency and is equivalent to the difference between the synchronous speed and the actual shaft speed of the machine.

The excitation from the stator windings causes the generator to act in much the same way as a basic squirrel cage or wound rotor generator,. Without the additional rotor excitation, the frequency of a slow running generator will be less than the grid frequency which provides its excitation and its slip would be positive. Conversely if it was running too fast the frequency would be too high and its slip would be negative.

The rotor absorbs power from the grid to speed up and delivers power to the grid in order to slow down. When the machine is running synchronously the frequency of the combined stator and rotor excitation matches the grid frequency, there is no slip and the machine will be synchronised with the grid.

- Grid Side Converter (GSC): Carries current at the grid frequency. It is an AC to DC converter circuit used to provide a regulated DC voltage to the inverter in the machine side converter (MSC). It is used maintain a constant DC link voltage. A capacitor is connected across the DC link between the two converters and acts as an energy storage unit. The grid side converter is used to maintain a constant DC link voltage. In the opposite direction the GSC invereter delivers power to the grid with the grid regulated frequency and voltage.

  As with the in-line converter described above, by adjusting the timing of the GSC inverter switching, the GSC converter also provides variable reactive power output to counterbalance the reactive power drawn from the grid enabling power factor correction as in the in-line frequency control system described above.

- Machine Side Converter (MSC): Carries current at slip frequency. It is an DC to AC inverter which is used to provide variable AC voltage and frequency to the rotor to control the torque and speed of the machine.

  When the generator is running too slowly, its frequency will be too low so that it is essentially motoring. The machine side converter takes DC power from the DC link and provides AC output power at the slip frequency to the rotor to eliminate its motoring slip and thus increase its speed. If the rotor is running too fast causing the generator frequency to be too high, the MSC extracts AC power from the rotor at the slip frequency causing it to slow down, reducing the generator slip, and converts the rotor output to DC passing it through the DC link to the GSC where it is converted to the fixed grid voltage and frequency and is inserted into the grid.

DFIG Control

Frequency

The frequency of the rotor currents induced by transformer action from the stator is the same as the slip frequency and this is equivalent to the frequency error signal in the feedback loop.

The additional direct excitation of the rotor adds a second set of controlled currents to the currents already induced in the rotor by transformer action from the stator. These additional currents affect the rotation speed of the rotor in the same way as the stator induced currents, producing an additional driving torque on the rotor except that the additional rotor currents are independent of the speed of the rotor. The frequency of the control current supplied by the MSC can be precisely controlled to match and thus neutralise the slip frequency so that, with zero slip, the generator rotates at the synchronous frequency determined by the grid. The greater the slip, the greater the compensating frequency required. The control system has to respond to both positive (motor) slip and negative (generator) slip.

To increase the speed of a slow running rotor, the phase sequence of the rotor windings is set so that the rotor magnetic field is in the same direction as the generator rotor producing negative slip to counteract and thus neutralise the rotor's positive slip. To reduce the rotor speed, the phase sequence of the rotor windings is set in opposite direction from the generator's rotation producing positive slip to counteract the rotor's negative slip.

When operating at synchronous speed the rotor current will be DC current and there will be no sip and no power flow through the rotor.

Voltage

The generator output voltage is determined by the magnitude of the excitation current supplied to the rotor and this can be adjusted by means of the rotor input voltage provided by the MSC. A chopper or pulse width modulator PWM is used to generate the variable DC control voltage necessary. The converter feedback controls thus enable the excitation current to be regulated by the MSC to neutralise the voltage error signal and thus obtain a constant bus voltage matched to the grid voltage.

DFIG Performance

- The DFIG system provides regulated power tied to the grid frequency and voltage when driven by varying levels of torque from the wind.
- Typical speed control range is ± 30% of synchronous speed.
- For a greater speed control range it may be necessary to implement separate pitch control on the wind turbine's rotor vanes.
- The generator power flow is shared by the stator and the rotor with 70% or more coming from the stator. The feedback loop only carries the slip power which is between 20% and 30% of the total.
- Because of the reduced power flowing through the converters, compared with the in-line control system described above above, they the DFIG converters can be implemented with less expensive lower power components.
- The DFIG machine can produce up to twice the power of a similar sized singly fed machine while incurring similar losses, however the losses in the electronic controls must be added to this. Nevertheless the DFIG machine efficiency is better that a singly fed machine.

Wind Farms

Grouping 10 to 100 wind turbines together in so called "wind farms" can lead to savings of 10% to 20% in construction, distribution and maintenance costs. According to NREL the"-footprint" of land needed to provide space for turbine towers, roads, and support structures is typically between 0.1 and 0.2 hectares (0.25 and 0.50 acres) per turbine. With the typical capacity of wind turbines installed in existing wind farms being around 2 MW, it would take a wind farm with 2000 wind turbines covering 200 to 400 hectares (500 to 1000 acres) just to replace the 4000 MWe power generated by the UK's Drax coal fired power station.

Unfortunately for the economics of wind turbines, the utility company needs to keep the equivalent capacity from other sources (conventional generating stations or batteries) just to keep the grid customers supplied when the wind is not blowing.

Domestic Wind Turbine Installations

1.6 kW Wind Turbine with 2.8 Metre Diameter Rotor.

In a typical domestic system the wind turbine is coupled directly to a three phase asynchronous permanent magnet AC generator mounted on the same shaft. To save on capital costs, domestic installations do not have variable pitch rotor blades so the rotor speed varies with the wind speed. The generator output voltage and frequency are proportional to the rotor speed and the current is proportional to the torque on the shaft. The output is rectified and fed through a buck-boost regulator to an inverter which generates the required fixed amplitude and frequency AC voltage.

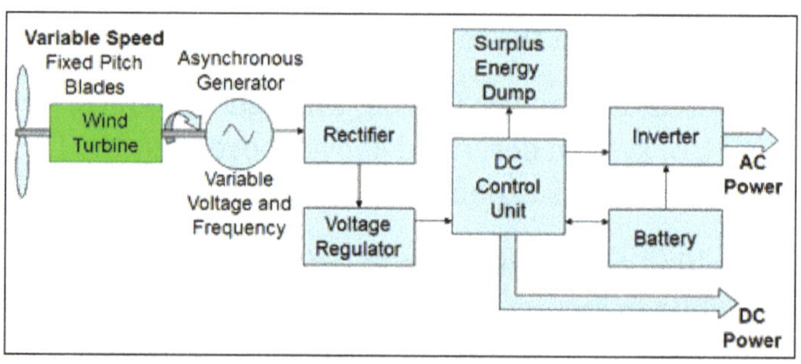

Small Scale Wind Power (Domestic Systems).

There is possible confusion in the classification of the generator. It is actually a synchronous generator because the frequency of its output is directly synchronised with the rotor speed. In this application however it is called an asynchronous generator because the output frequency of the generator is not synchronised with the mains/utility frequency.

Urban Installations

Wind turbine blade sizes in urban applications are usually limited for practical reasons to less than about 1 metre (2 metres diameter) as well as by local planning ordinances and for similar reasons the height of the turbine above ground is limited to just above rooftop level but below treetop level.

1. Economics:

A typical domestic installation with a 1.75m swept diameter, (swept area of 2.4m^2), costs around £1500 ($2250). At the rated wind speed of 12.5m/s (28 mph) the wind power intercepted will be 2870 Watts, but after taking into account all the unavoidable system losses, the actual electrical output power will be around 1000 Watts. However this is at the upper end of the performance possibilities. Wind turbulence and shielding due to buildings and trees inhibits sustained strong, gust free wind flow and in any case, for most of the time, the wind speed will more likely be towards the lower end of the performance specification at 4 m/s (9 mph), that is a light breeze. At this speed the power output of the system will be about 32 Watts - not enough to power a single light bulb. For much of the time the power generated could be less than the quiescent power drain of the inverter.

Running with a constant power output of 32 Watts for a full year would generate only 280 kWh (280 Units) of electrical energy worth £28 at today's price of £0.10 ($0.15) per kWh. To put it into perspective, a typical UK household consumes about 5,000 kWh of electrical energy per year.

Because the system is connected directly to the grid there is no need for battery back-up and in any case the cost of the batteries would make an already weak economic case for the system even weaker.

Thus small domestic rooftop wind turbine installations do not make a serious contribution to the household energy supply.

Self-sufficiency and selling surplus energy back to the utility are out of the question and the payback period on the capital investment is out of sight.

2. Carbon Footprints:

As with solar power, if the investment fails the conventional economic tests, the notion of carbon footprints is often used to justify the expense, based on the potential for reducing the amount of greenhouse gases emitted by alternative methods of power generation.

Rural Installations

The economics of rural and remote locations make wind power more attractive than for urban locations. Because of the remoteness, connection to the electricity grid may be impossible or prohibitively expensive. Furthermore, larger, more efficient wind power installations are possible and the prevailing winds will also be higher.

Hybrid Installations

Hybrid systems combining wind and solar power provide energy diversity reducing the risk of power outages. Wind speeds are often high in the winter when the available solar energy is low and low in the summer when the available solar energy is high.

Wind Energy Conversion Systems

The wind energy conversion system (WECS) includes wind turbines, generators, control system, interconnection apparatus. Wind Turbines are mainly classified into horizontal axis wind turbines (HAWT) and vertical axis wind turbines (VAWT). Modern wind turbines use HAWT with two or three blades and operate either downwind or upwind configuration. This HAWT can be designed for a constant speed application or for the variable speed operation. Among these two types variable speed wind turbine has high efficiency with reduced mechanical stress and less noise. Variable speed turbines produce more power than constant speed type, comparatively, but it needs sophisticated power converters, control equipments to provide fixed frequency and constant power factor.

The generators used for the wind energy conversion system mostly of either doubly fed induction generator (DFIG) or permanent magnet synchronous generator (PMSG) type. DFIG have windings on both stationary and rotating parts, where both windings transfer significant power between shaft and grid. In DFIG, the converters have to process only about 25-30 percent of total generated power (rotor power connected to grid through converter) and the rest being fed to grid directly from stator. Whereas, converter used in PMSG has to process 100 percent power generated, where 100 percent refers to the standard WECS equipment with three stage gear box in DFIG. Majority of wind turbine manufacturers utilize DFIG for their WECS due to the advantage in terms of cost, weight and size. But the reliability associated with gearbox, the slip rings and brushes in DFIG is unsuitable for certain applications. PMSG does not need a gear box and hence, it has high efficiency with less maintenance. The PMSG drives achieve very high torque at low speeds with less noise and require no external excitation. In the present trend WECS with multibrid concept is interesting and offers the same advantage for large systems in future. Multibrid is a technology where generator, gearbox, main shaft and shaft bearing are all integrated within a common housing. This concept allows reduce in weight and size of generators combined with the gear box technology. The generators with multibrid concept become cheaper and more reliable than that of the standard one, but it loses its efficiency.

To achieve high efficient energy conversion on these drives different control strategies can be implemented like direct torque control (DTC), field oriented control (FOC). The FOC using PI controller has linear regulation and the tuning becomes easier. The wind turbine electrical and mechanical parts are mostly linear and modeling will be easier. The blade aerodynamics of the wind turbine is a nonlinear one and hence the overall system model will become nonlinear. The wind energy conversion system which will be modeled as shown in figure may not be optimal for extracting maximum energy from the resource and hence various optimization techniques are used to achieve the goal.

WECS Modeling

The basic device in the wind energy conversion system is the wind turbine which transfers the kinetic energy into a mechanical energy. The wind turbine is connected to the electrical generator through a coupling device gear train. The output of the generator is given to the electrical grid by employing a proper controller to avoid the disturbances and to protect the system or network. Figure below shows the overall block diagram of the wind energy conversion system (WECS). Here, $V\omega$ represents wind speed, P_ω, P_m and P_e represent wind power, mechanical power and electrical power respectively.

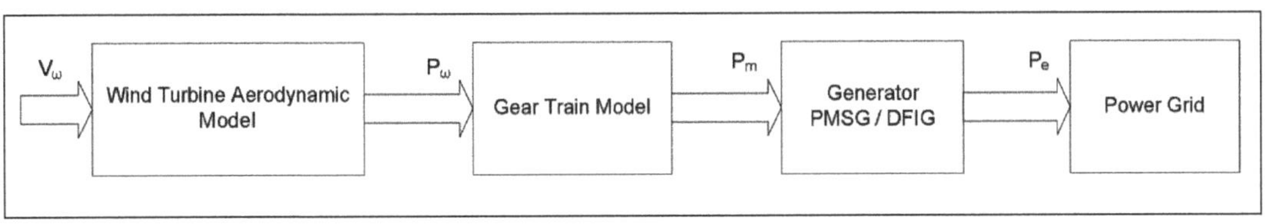

WECS Block Diagram.

Wind Turbine

Wind energy is transformed into mechanical power through wind turbine and hence it is converted into electrical power. The mechanical power is calculated by using the following equation,

$$P_m = 0.5\,\rho A C_p(\lambda, \beta) v_{wind}$$

where, ρ is the air density which normally takes the value in the range 1.22 – 1.3 kg/m^3, A is the area swept out by turbine blades (m^2), v_{wind} is the wind speed (m/s), $C_p(\lambda, \beta)$ is the power coefficient which depends on two factors: β, the blade pitch angle and the tip speed ratio and, λ, which is defined as:

$$\lambda = \Omega.R / v_{wind}$$

where, Ω is the angular speed (m/s) and R is the blade radius (m). The power coefficient, C_p is defined as,

$$C_p(\lambda,\beta)=C_1\left(\frac{C_2}{\lambda_i}-C_3\beta-C_4\right)\exp\left(\frac{-C_5}{\lambda_i}\right)+C_6\lambda$$

Where,

$$\frac{1}{\lambda_i}=\left(\frac{1}{\lambda+0.08\,\beta}-\frac{0.035}{\beta^3+1}\right)$$

and coefficients C_1 = 0.5176, C_2 = 116, C_3 = 0.4, C_4 = 5, C_5 = 21, and C6 = 0.0068.

The power coefficient is nonlinear, and it depends upon turbine blade aerodynamics and it can be represented as a function of tip speed ratio, λ. The optimum value of λ corresponds to maximum of C_p from the power coefficient-tip speed ratio curve.

Figure below shows the power coefficient with respect to tip speed ratio. It is observed that the maximum power coefficient value $C_{p\_max}(\lambda, \beta) = 0.48$ for λ = 12 and for β = 0°. This particular value of λ_{opt} results in optimal efficiency point where maximum power is captured from wind by the turbine.

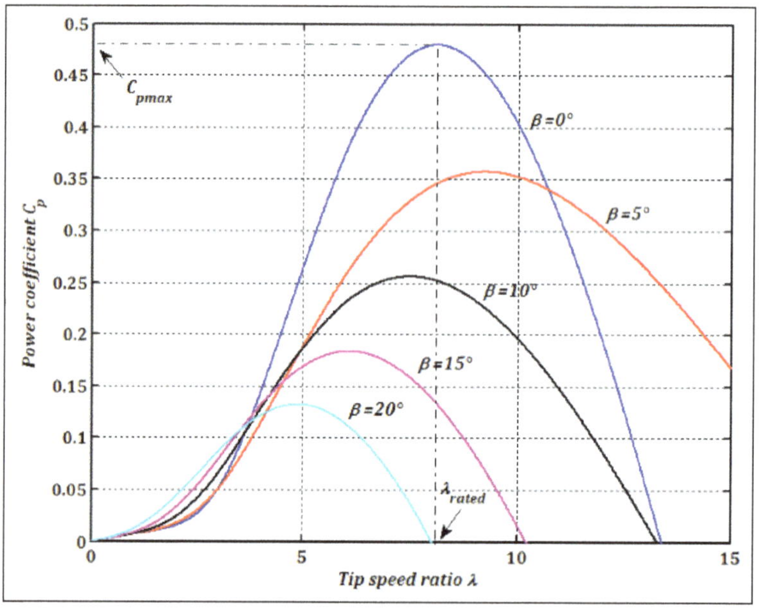

Power Coefficient of the Wind Turbine Model.

As the power transmitted is assumed to be the product of rotational speed with mechanical torque, the rotational torque is obtained as:

$$T_m = P_m/\Omega$$

Thus the optimal angular speed is achieved through the relation,

$$\Omega_{opt} = \lambda_{opt}\, v_{wind} / R$$

and the maximum mechanical power is,

$$P_{m\_max} = 0.5\,\rho A C_{p\max}\, v^3_{wind}$$

Figure below shows the wind turbine power characteristics obtained for various values of the wind tangential speed. Here it can be observed that maximum power (active) is achieved through optimal wind speeds and not at high wind velocity. The wind turbine does not operate when the wind speed is less than the minimum speed because the captured wind energy is not enough to compensate the losses and operation cost.

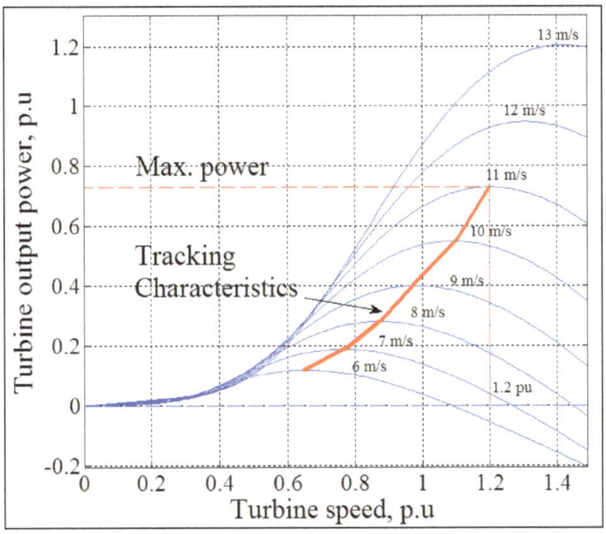

Wind Turbine Power Characteristics.

Drive Train Modeling

The gear box enables to convert the slow speed of the turbine blades to high speed required by the electrical machine through the gear box ratio, g_r. This element converts the mechanical torque and the machine speed as follows:

$$T_{aero} = g_r . T_m$$
$$\Omega_m = g_r\ .\Omega$$

where, T_{aero} is the aerodynamic torque and Ω_m is the machine speed. The drive train is modeled by the simplified motion equation as follows:

$$J\frac{d\Omega_m}{dt} = T_m - T_e f \Omega_m$$

where J is the mechanical inertia of the wind turbine and generator, T_e is the electromagnetic torque, and f is the friction coefficient.

PMSG Modeling

The modeling of PMSG type electrical equipment is made through the following equations, represented by d-q reference frame:

$$V_d = R_s i_d + L_d \frac{di_d}{dt} - \omega_e L_q i_q$$

$$V_q = R_s i_q + L_q \frac{di_q}{dt} - \omega_e L_d i_d + \omega_e \varphi_m$$

Where, V_d and V_q are d and q components of stator voltages (V), i_d and i_q are d and q components of stator currents (A), R_s is stator resistance (ohms), L_d and L_q are machine inductances (H), ω_e is the electrical speed (rad/s) and φ_m is the magnetic flux (wb). The electrical torque is obtained through the following equation:

$$T_e = \frac{3}{2} p \left\{ \varphi_m i_q + \left(L_d - L_q \right) i_d i_q \right\}$$

Where, p is the pair of poles. The rotor dynamics of the machine is given by,

$$T_m - T_e = B\,\omega_r + J \frac{d\omega_r}{dt}$$

Where, B is the rotor friction (kgm²/s), J is the rotor inertia (kgm²), ω_r is rotor speed (rad/s) and T_m is the mechanical torque produced by wind (Nm). The machine dynamics can be simplified by assuming ($L_d = L_q = 0$) and the dreference current is zero ($i_d^* = 0$) and hence the product term $(L_d - L_q)\, i_d\, i_q$ is negligible.

DFIG Modeling

The commonly used model for induction generator is the Park model. To obtain its state model we neglect stator resistance, and then we have:

$$\Phi_{sq} = 0, \frac{d\Phi_{sd}}{dt} = 0, V_{sd} = 0 \text{ and } V_{sq} = -V_s$$

Where, V_s is the stator voltage. The reduced state model is obtained as follows:

$$V_{rd} = \sigma L_r \frac{di_{rd}}{dt} + r_r i_{rd} - (\omega_s - \Omega)\sigma L_r i_{rq}$$

$$V_{üüüüüü} = \sigma L \frac{di_{rq}}{dt} + r\, i + (\omega - \Omega)\sigma L\, i + \frac{M}{L_s}\Phi\ (\omega - \Omega)$$

Where, V_{rd} and V_{rq} are the d and q components of rotor voltage (V), V_{sd} and V_{sq} are the d and q components of stator voltage (V), i_{rd} and i_{rq} are the d and q components of rotor current (A), Φ_{sd} and Φ_{sq} are the d and q components of stator flux (wb), ω_s is the synchronous pulsation, Ω is the generator speed, r_r is the rotor resistance, L_s and L_r are the stator and rotor inductances (H), M is the mutual inductance (H) and σ is the leakage parameter given as, $\sigma = 1 - (M^2 / L_s L_r)$.

DFIG is one of the most important generators and are widely used for variable speed WECS. Nowadays, this type of DFIG-WECS has a part of wind energy market, which is close to 50%. Major wind turbine producers manufacture wind energy conversion system based on DFIG's but the difficulties associated in complying with grid side ride-through requirement may limit its use in future. To overcome the above stated difficulty, PMSG technology looks most promising in WECS, which shares nearly 45% in wind energy market. Multi-pole PMSG's with full power back-to-back converters appears to be the configuration to be adapted by most of the wind turbine manufacturers in near future, gradually replacing DFIG in wind energy market. Wind energy conversion systems with the range of 1.5 MW to 3 MW are manufactured by 44 companies in 163 models. Among these 60 models uses DFIG, 66 models uses PMSG, 18 models uses cage type IG and 19 models uses synchronous generator with external excitation.

Control Strategies

DFIG-WECS Control

Controller is designed to adjust the turbine speed to extract the maximum power from wind source. Usually PI controller is designed for this purpose according to the estimated parameters either offline or online and the observed disturbance torque is feed forward to increase system robustness. But this type of classical controller is not enough to serve the purpose efficiently. Hence various improvements are made for the controller to achieve the requirement.

The general structure of control block diagram in the DFIG-WECS having two levels of control. The lower level control being the electrical control system, i.e. torque and reactive power control. The rotor side controller (RSC) and grid side controller (GSC) are employed to serve the electrical control system. To compromise the dynamics of variable speed WECS Poitiers et al. has implemented a two degree freedom Regulation-Solution-Tracking (R-S-T) controller, which is a widely preferred digital controller and Linear Quadratic Gaussian (LQG) controller with state feedback. For the ease of implementation of controller tuning the LQG controller is preferred than R-S-T controller.

To reduce the output power fluctuations of DFIG with a flywheel energy storage system (FESS) Jerbi et al. has proposed a fuzzy supervisor using sugeno fuzzy model based on two parameters flywheel speed and wind power accessibility so as to ensure a smooth reactive power to the load supplied by the wind generator. This ensures to reduce the voltage fluctuations. For the variable speed WECS has proposed a method which will combine fuzzy neural network and sliding mode speed observer. This technique allows faster convergence to a simple linear dynamic behavior, even in presence of parameter changes and uncertainties which are the major problems expected from variable speed drive control. As most of the controllers depends on torque control M. Pucci proposes a method based on speed control of the machine. This is achieved through total least squares (TLS) EXIN full order observer which turns as an intelligent sensor less technique for wind generation control.

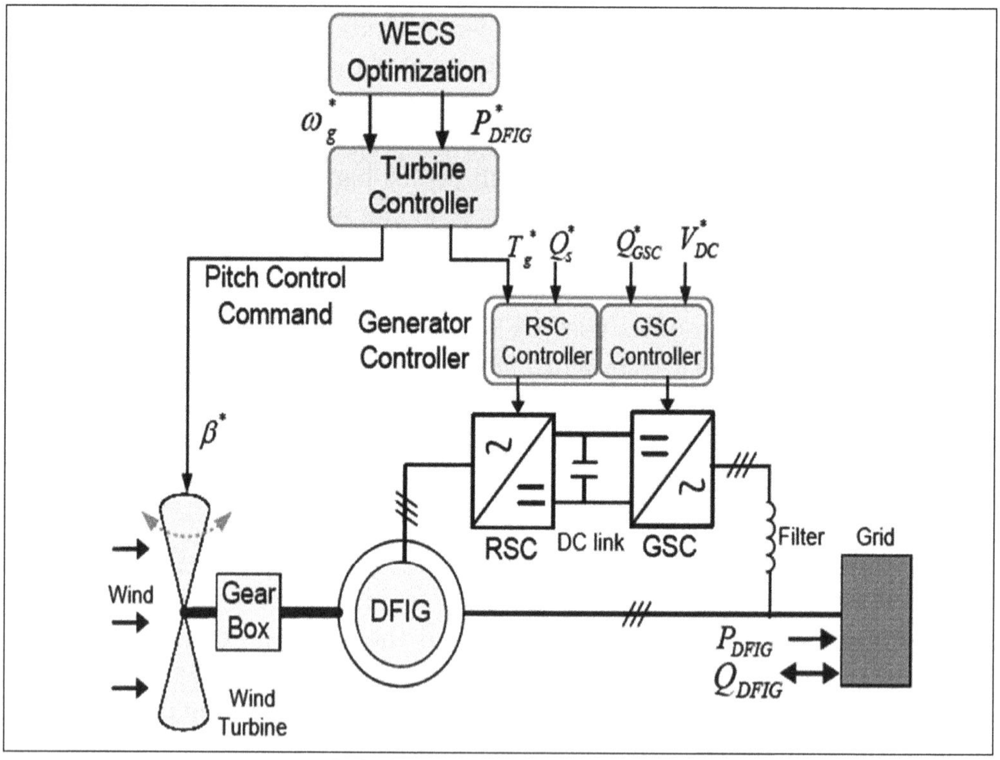

Control Block Diagram of DFIG Based WECS.

There are other types of controllers using PID for controlling the WECS with optimal power output. Muhando et.al proposes a self-tuning regulator by incorporating recursive least square algorithm to predict the process parameters and update the states. This regulator gives more efficient compared to classical PID controller. To achieve optimal power, adaptive fuzzy PID control strategies are proposed and proved the performance is improved than that of conventional one.

Many research papers discuss the intelligent controllers with sensor less wind energy control. Whei-Min Lin et al. proposed a technique where design is made online through recurrent fuzzy neural controller (RFNN) with high performance model reference adaptive system (MRAS) observer which replaces the use of sensors to measure the parameters. Lin et.al suggested that to increase the learning ability of back propagation network of RFNN modified particle swarm optimization is used (mPSO). This provides fast and accurate velocity information to avoid anemometers and rotor position is estimated from flux linkages.

Behra and Rao investigated novel grid side controller, modeled in synchronously rotating reference frame to improve the dynamic response on various load conditions. Liu and Hsu discuss the stator resistance effect on the excitation voltage, which gives the maximum output power and minimal loss for the DFIG. The DFIG-WECS model can be validated through various simulation tools. To improve the real time simulation time FPGA's can be incorporated using Runge-Kutta numerical integration algorithm in compared to PC based simulation.

PMSG-WECS Control

Figure below shows the block diagram of PMSG based WECS with two stages as optimization and electrical controllers. The various techniques are discussed as below.

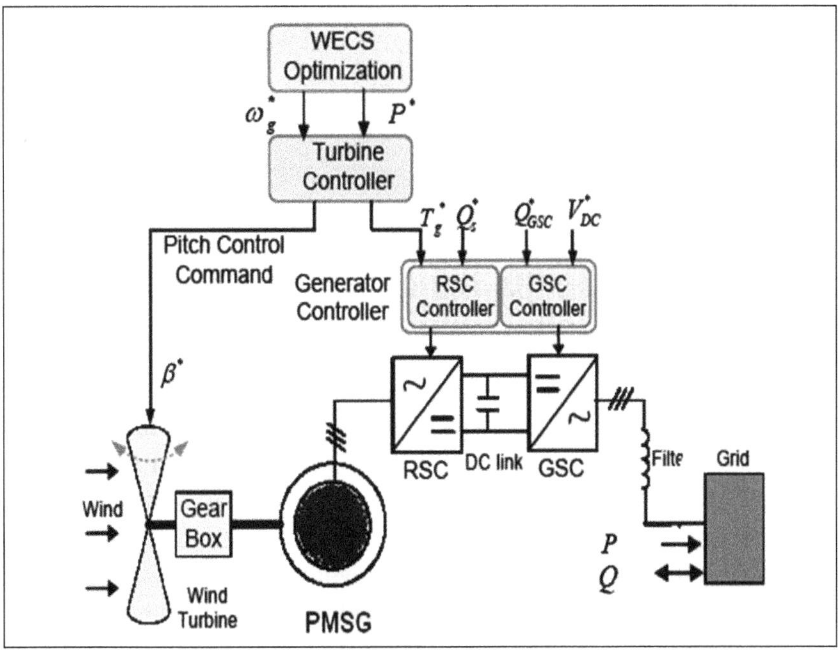

Control Block Diagram of PMSG Based WECS.

Jordi Zaragoza et al. discusses the control of PMSG based WECS using field oriented control (FOC) for controlling speed regulation and generating current. A PI controller is used for this purpose and the tuning parameters are determined through zero-pole cancellation (ZPC) strategy. Jemaa Brahmi et.al compares three control methods for PMSG based WECS. In the first model MRAS observer is used to estimate the parameters of the controller by comparing output of reference model and adaptive model. Secondly artificial neural network (ANN) based observer were trained to produce desired output correction to the estimated speed. And lastly, sliding mode observer (SMO) based control is discussed. Among these three techniques SMO and ANN observer gives good results when static resistance varies and chattering phenomenon is observed with SMO. The static resistance causes a static error in case of MRAS observer. The SMO is more robust than the other two methods.

Kelvin Tan and Syed Islam proposed a sensor less optimal control strategy using fuzzy power mapping technique. Using the result of power mapping loop and alternator frequency derivative loop, the controller allows the bus voltage to vary to maximum power operating point. Bao et al. proposes an adaptive control strategy for variable speed wind turbine using neural based identifier and inverse model controller which has good learning capability and high performance than nonlinear controller. Wang et.al proposes indirect matrix converter of PID type which has more degree of freedom and can be implemented in an easy way for the PMSG based WECS. This shows excellent dynamic performances with lower mechanical stress and robust in control.

There are other techniques for controlling the WECS. Wind prediction control scheme is the one where the control parameters are obtained through forecasting the wind using autoregressive statistical models. This method will avoid the mechanical sensors used to get the actual parameter tracking.

Optimization of WECS

The next challenging task in WECS is to operate the system at the optimal point so as to increase the output. The optimization blockand for DFIG and PMSG based WECS. Among the various techniques followed for this, maximum power point tracking (MPPT) plays a vital role.

Kesraoui et al. proposed the MPPT algorithm using step and search method to extract the maximum power from the wind turbine from cut-in to rated wind velocity by sensing the dc link power. Iulian Munteanu et al. used wind turbulence as a search for MPPT instead of sinusoidal search. Here high speed shaft's average speed is slowly adjusted using FFT of measures from system as an estimate to the optimal point. This method is effective when the wind turbulence level is high. The same author also proposes sliding mode control, optimal regime tracking, and frequency separation principle for optimizing the WECS based on PMSG type. Syed Muhammad et.al proposes a novel method of MPPT along with Hill Climb Search (HCS) algorithm. This method performs self-tuning to cope with the non-constant efficiency of the generator- converter systems.

Adaptive neuro fuzzy inference system (ANFIS) based MPPT is discussed in Meharrar et al. Here variation of wind speed is used as input for ANFIS to predict optimal speed rotation and back propagation learning algorithm is used for training this network. This method shows better response than compared to FLC. Other type of fuzzy controller was proposed by Calderaro et al. using Takagi-Sugeno-Kang (TSK) model to maximize energy extraction which avoids sensors in MPPT. To estimate the optimum value for generator support vector regression (SVR) algorithm is used by Lee et al. It is proved that the generator losses are reduced up to 40% and also excellent in accuracy. Kazmi et al. proposes growing neural gas (GNG) based MPPT which is having FOC integrated with TLS EXIN observer without sensors to optimize the system.

A new method of MPPT which uses gradient approximation algorithm is proposed by Ying-Yi Hong et al. This method does not require anemometer, tachometer and information of wind turbine generator system characteristics. As the gradient approximation converges very fast and the implementation is easy, this method is efficient for an optimization problem. Kongnam and Nuchprayoon propose to apply PSO to solve for optimum rotor speed in fixed and variable speed operation of WECS. In this power and energy are more dependent of mean wind speed than weibull distribution of speed.

Pitch angle control of wind turbine turns out to be another optimization problem in WECS. Yilmaz and Zafer discusses radial basis function based neural network controller which tracks the reference signal based on training to achieve better controller. Boukhezzar et al. propose control strategy by combining torque control and pitch control which leads to good performance in rotor speed regulation and electrical power regulation with acceptable control loads compared to classical PID and LQG controllers.

Windmill

A windmill is a structure used to harness the power of the wind for purposes like grinding grain, pumping water, and generating electricity. The power of the wind was first harnessed by sailors, who were able to understand lift and harness the winds power through sails. This knowledge led to the development of the first vertical axis sail-type windmill used by the ancient Persians and Chinese for grinding grain and pumping water.

They consisted of vanes called sails or blades that when prompted to turn by the wind, converted the wind's energy into rotational energy that could be utilized. Early European windmills with horizontal axis systems were the foundation for current wind turbine technology used for energy production.

Modern Advances

Mechanical water pumping mills were first developed in the United States in 1854. These mills originally consisted of four wooden blades, with steel blades coming into existence in 1870. More than six million mechanical wind mills were installed in the United States between 1850 and 1970. Their main applications were for stock watering and water for farm houses. Very large windmills were used to pump water for steam trains.

The first windmill to generate electricity was invented in 1888 by an Ohio-based engineer named Charles Brush. The windmill could generate up to 12 kilowatts of electricity and was a sixty foot post mill with a 56 foot rotor. In today's society windmills used to generate electricity are referred to as wind turbines.

The wind turbine market evolved from the 1 to 25 kilowatt range for domestic and agricultural practices, to intermediate-scale machines in the 50 to 600 kilowatt range for interconnected utility wind farm applications between the years of 1973 and 1986.

Over this time period, more than 20 different designs where engineered and tested, most proving to be unfeasible and inefficient. Wind turbines evolved from the four bladed systems first created in the United States, to two bladed systems, and finally to the three bladed systems commonly seen in wind farms today.

How Windmill Works

The earliest documented windmills were vertical axis systems, with vertical sails made from reeds or wood that was attached by horizontal struts to a vertical shaft with a stone for grinding purposes. The first windmill to appear in Europe was the post mill with a horizontal axis system, with a four-bladed mill attached to a central post.

Mostert's Mill describes the post mill as a building balanced and pivoted on a vertical central post kept upright by crosstrees and angled quarter bars, whose ends carried four upright stone or brick pillars.

The mill consisted of a wallower, a large brake wheel located on the same shaft as the sails, which transferred power to a smaller gear. The wallower shared the vertical shaft with the great spur wheel, from which a stone nut drove the grinding stone.

The tower mill was an advancement of the post mill having multiple floors for storing grain, removing chaff, grinding grain, and living quarters for the windsmith and his family.

The most important feature of the tower mill was a cap (roof) that could pivot in response to changing wind patterns. Unlike the post mill where you had to move the entire structure to orient the mill, only the cap had to be moved in the tower mill.

Both the tower and post mills were originally designed to be manually oriented into the wind by pushing a lever located on the back of the mill. In 1745, Edmund Lee invented the windmill fan tail, which was mounted on the rear of the mill at right angles to the sails automatically turning the cap to bring the sails into the wind.

Smock mills are similar to tower mills differing mainly in their appearance, in which smock mills were octagonal or hexagonal with their six to eight sides, rather than circular. Smock mills have the same rotating cap design as the tower mills, but were typically much larger.

The first mechanical mills that appeared in the United States had four wooden blades that resembled paddles, and most were equipped with "tails" that oriented them into the wind. However, some of these mechanical mills were designed to operate downwind, called weather-vaning mills.

Some mechanical mills sported speed control provided by hinged blades that folded back in high winds reducing the thrust, by reducing the rotor capture area. When steel blades where introduced they required a reduction gear to compensate for their high speed, so that the standard reciprocal pumps were at the right speed for the mill to function properly.

The three-bladed turbine systems commonly seen today, function in much the same way as the original post mills. Wind turns two or three blades around a rotor that is connected to a main shaft that spins a generator to create electricity.

The rotor, which is the blades and hub combined, are mounted 100 feet or more above the ground to take advantage of faster, less turbulent wind. The rotor is attached to a pitch system that turns the blades out of the wind to control the rotor speed. When winds are too high or too low to produce electricity the pitch system prevents the rotor from turning.

The pitch system is connected to a low speed shaft that turns at about 30-60 rotations per minute. The low speed shaft is connected to the gear box which turns the low rotational speeds into the high rotational speeds of 100-1800 rpms required by the generator, which is attached to the gearbox, to produce electricity.

Today's turbines also consist of brake system that can mechanically, electrically, or hydraulically stop the rotor in emergencies, and a controller that starts the machine up at wind speeds that are between 6 to 16 mph and shuts the machine off when wind speed are above 55 mph to prevent damage. An anemometer measures the wind speed and transmits the data to the controller. The electricity is then transferred down the tower and converted to the same voltage as the local energy grid.

Wind Farm

A wind farm is any group of wind turbines that, together, generate a sizeable amount of electricity and act as a power plant. Individual wind turbines were used to generate electricity as far back as 1887. Professor James Blyth of Glasgow constructed a 10m windmill that charged an early battery, which he used to power the lights in his home – although those living in the local town of Marykirk turned him down when he offered to light the main street from this windmill, saying that "electrical power was the work of the devil." Devil's work or not, we all rely on it to an increasing extent in the modern world.

Wind Turbines

It took nearly a century from Professor Blyth's windmill for the modern, industrial-scale harvesting of wind energy to take root. The first modern wind farm was constructed in 1980, in New England. Like so much renewable energy research and development, it was sparked by the oil/energy crisis of 1973, which led the Carter administration to invest in solar and wind power as alternatives to dependence on foreign oil. This wind farm used 20 small wind turbines to generate 0.7 MW of power; they've grown in size since then.

Large wind farms can consist of hundreds of turbines spread over a large area of land; the largest wind farm in the world is the Gansu wind farm in China. The aim was to have 20 GW of power capacity installed by 2020; currently, 6.5 GW has been installed – over 7,000 wind turbines. The winds that sweep in over the Gobi desert have proved fertile ground for this new kind of farm.

Various estimates have been made for the global potential for wind power, with the most optimistic suggesting that we could generate many times more electricity than the world currently consumes using a combination of onshore and offshore wind farms. Lu et al, for example, suggested that wind turbines could provide more than 40x current global electricity consumption – and several times our global energy consumption in all forms. However, finding the appropriate location for wind farms has proved a difficult challenge.

Naturally, you need wide-open space, although the area between turbines can be used for certain kind of agriculture (typically grazing land.) At the same time, you need an average wind-speed of greater than around 4.5 m/s to be economical – the faster the windspeed, the quicker the project will pay for itself and begin realising profit, so this is an important factor.

Consistent windspeeds throughout the year are ideal for the stability of the grid, and it's usually considered advantageous to build the windfarms at height or in flat regions where the windspeed isn't disrupted by orography or man-made structures. Some regions can provide particularly good windspeeds – an example is 'micrositing', where wind turbines are placed on ridgelines to take advantage of the acceleration of the wind due to the sudden change in topography.

These requirements must be balanced with the need to build the wind farm reasonably close to places that need a power supply, such as cities. If the electricity has to be carried a great distance, even with high-voltage DC cables, the losses can render the power plant less economical than it would be otherwise.

However this consideration has to be balanced with the prospect that some people find the wind turbines unsightly in the countryside – a phenomenon called NIMBYism. Typically site selection for wind farms requires consideration of all of these factors, followed by a measurement campaign to narrow down between sites that meteorological data suggests will be useful.

Offshore Wind Farms

A more recent phenomenon has been the creation of offshore wind farms. The first of these was installed in Denmark in 1991, and they have grown in stature since. Opening up the ocean wind resource to exploitation increases the area that wind turbines can cover. It's also the case that wind can be substantially stronger offshore; driven by the heating of the oceans during the day, that sea breeze can be strongest in the afternoon, when power demand is increased.

Typically it's considered best to locate them just offshore from major coastal settlements, reducing the length of electrical cable and associated losses. Until 2017, most wind farms were fixed-foundation – this required them to be built in shallow waters less than around 30 m in depth, allowing bases for the wind columns to extend down to the sea floor. Renewable energy fans will be interested to watch the progress of the world's first floating wind farm, which could potentially allow wind potential from the deep ocean to be harnessed.

The floating wind farm, consisting of five 175 m turbines, started supplying power. Early indications are that it is operating at high capacity, and, with further development, it may prove to be a major player in the coming energy transition.

Offshore Wind Power

Offshore wind power has the potential to generate up to twice as much power as onshore. Offshore wind turbines harness the power of sea winds to generate carbon-free renewable energy. They work in exactly the same way as onshore generators; the only difference being that it is installed at sea and is either anchored to the seabed or mounted on a floating structure. Using masts much higher than those of onshore turbines, offshore wind generators can generate up to twice as much energy than their onshore counterparts, because they can rely on higher and more consistent wind speeds.

Last year, offshore wind farms achieved a new level of total generating capacity, with more than 4,000 MW worldwide. The majority of this growth came from two new projects; one in Germany (2,400 MW) and the other in China (500 MW). These two countries lead the industry alongside the UK, which still boasts the largest offshore wind generating capacity (5,400 MW of installed capacity at the end of 2015).

Floating Wind Turbines: Generating Power in the Deep Offshore

Like fixed offshore wind turbines, the floating wind turbine generates electricity by converting the mechanical energy of the wind. But its floating structure enables it to generate power a long way offshore in deep water where the sea winds are stronger and more consistent. Another benefit is its ease of installation, since there is no need to cast solid foundations or to use the kind of specialist construction vessels needed for fixed offshore wind turbines. At between 15 and 22 km from the coast, the visual and maritime impact of these windfarms is also relatively low.

Floating wind power technology may still be at the pilot plant stage, but this sector is of strategic importance for coastal regions in terms of jobs, industry and ecology. The technically exploitable potential of floating wind power in Europe is estimated at 600 GW, compared with 250 GW for fixed offshore wind power.

Onshore Wind Energy

Onshore wind refers to turbines located on land, while offshore turbines are located out at sea or in freshwater. Onshore wind plays a leading role in the generation of renewable electricity in the UK. In the third quarter of 2017 it generated 5.6 terawatt-hours (TWh) of electricity, an increase of 20 per cent on the previous year. This accounted for a quarter of the electricity generated by renewable sources in in that period, and of all electricity generation. In comparison, offshore wind produced 4 TWh over the same period. Together, onshore and offshore wind contributed nearly of total global renewable energy generation.

Main Advantages of Onshore Wind Energy

Wind energy has a relatively small carbon footprint. Some greenhouse gas emissions are created by the manufacture, transportation and installation of wind turbines, but these are considered fairly low, at around 9 gCO_2/KWh. By comparison, the average footprint of gas power generators is around 450 gCO_2/KWh and that of coal power generators around 1,050 gCO_2/KWh. Researchers found that wind power in the UK prevented the creation of almost 36 million tonnes of emissions from sources such as coal and gas between 2008 and 2014, the equivalent of taking 2.3 million cars off the road.

New onshore wind was identified in 2015 by Bloomberg New Energy Finance as being fully cost-competitive with gas and coal in some parts of the world and the cheapest source of power generation in the UK for the first time. This competitiveness was 'helped by cheaper technology but also by lower finance costs', while coal and gas became more expensive.

Main Disadvantages of Onshore Wind Energy

Onshore wind is an intermittent source of energy, as turbines cannot generate electricity on demand, but only when the wind is blowing, and at sufficient strength. When wind strength is insufficient for turbines to operate, fossil-fuel-based power supply is needed as backup, which can temporarily increase greenhouse gas emissions. The proportion of intermittent renewable electricity in the UK market is currently small enough to require very little backup, but as the share increases additional backup will be needed. However, other technologies, such as inter-linkages with other countries' grids, energy storage and electricity demand management, are expected to help tackle intermittency in the future, so the overall future impact on emissions is considered relatively low.

Onshore wind turbines have been criticised for their visual impact. Turbines are typically more spread out than other large-scale energy infrastructure projects and so can affect a larger area.

Another criticism is that species such as birds and bats may be affected by turbines. However, bird fatalities due to turbine collisions are relatively low compared with those from traffic or domestic cats and the benefits for wildlife of mitigating climate change are considered by conservation charities such as the RSPB to outweigh the risks, provided that the right planning safeguards are put in place, including careful site selection.

Airborne Wind Energy Systems

(Airborne Wind Energy) is usually employed to designate the high-altitude wind energy resource as well as the technological sector. High-altitude winds have been studied since decades by meteorologists, climatologists and by researchers in the field of environmental science even though many questions are still unsolved. The first work aimed at evaluating the potential of AWE as a renewable energy resource has been presented by Archer and Caldeira.

More in depth studies have been conducted employing complex climate models, which predict consequences associated with the introduction of wind energy harvesters (near surface and at high altitude), that exerts distributed drag forces against wind flows. Marvel et al. estimate a maximum of 400 TW and 1800 TW of kinetic power that could be extracted from winds that blow, respectively, near-surface (harvested with traditional wind turbines) and through the whole atmospheric layer (harvested with both traditional turbines and high altitude wind energy converters). Even if severe/undesirable changes could affect the global climate in the case of such a massive extraction, the authors show that the extraction of 'only' 18 TW (i.e., a quantity comparable with the actual world power demand) does not produce significant effects at global scale. This means that, from the geophysical point of view, very large quantity of power can be extracted from wind at different altitudes.

A more skeptical view on high altitude winds is provided in Miller et al. who evaluated in 7.5 TW the maximum sustainable global power extraction. But their analysis is solely focused on jet stream winds (i.e. only at very high altitude between 6 km and 15 km above the ground).

Despite the large variability and the level of uncertainty of these results and forecasts, it is possible to conclude that an important share of the worldwide primary energy could be potentially extracted from high altitude winds. This makes it possible to envisage great business and research opportunities for the next years in the field of Airborne Wind Energy.

Classifications of Airborne Wind Energy Systems

(Airborne Wind Energy Systems) is used to identify the whole electro-mechanical machines that transform the kinetic energy of wind into electrical energy. AWESs are generally made of two main components, a ground system and at least one aircraft that are mechanically connected (in some cases also electrically connected) by ropes (often referred to as tethers). Among the different AWES concepts, we can distinguish Ground-Gen systems in which the conversion of mechanical energy into electrical energy takes place on the ground and Fly-Gen systems in which such conversion is done on the aircraft.

In a Ground-Gen AWES (GG-AWES), electrical energy is produced on the ground by mechanical work done by traction force, transmitted from the aircraft to the ground system through one or more ropes, which produce the motion of an electrical generator. Among GG-AWESs we can distinguish between fixed-ground-station devices, where the ground station is fixed to the ground and moving-ground-station systems, where the ground station is a moving vehicle.

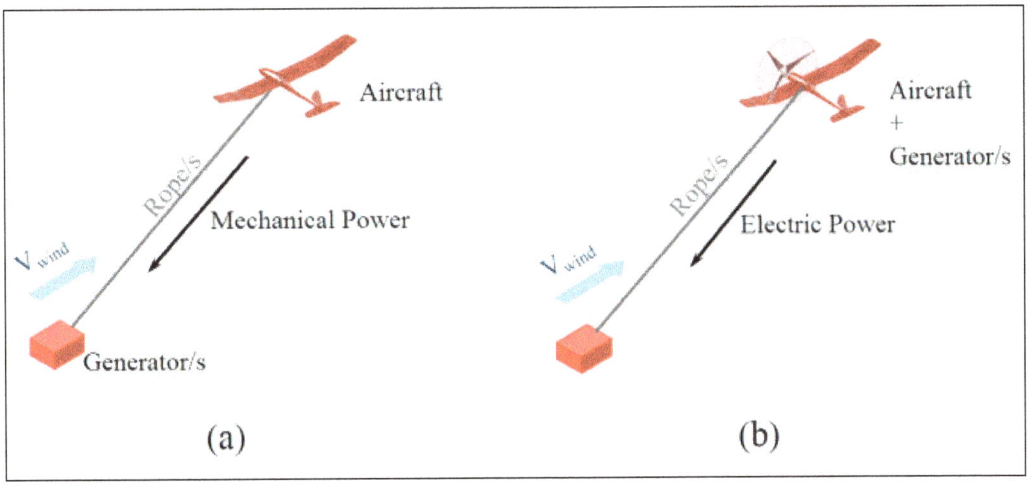

AWESs. Example of Ground-Gen (a) and Fly-Gen (b) AWESs.

In a Fly-Gen AWES (FG-AWES), electrical energy is produced on the aircraft and it is transmitted to the ground via a special rope which carries electrical cables. In this case, electrical energy conversion is generally achieved using wind turbines. FG-AWESs produce electric power continuously while in operation except during take-off and landing maneuvers in which energy is consumed. Among FG-AWESs it is possible to find crosswind systems and non-crosswind systems depending on how they generate energy.

Ground-Gen Airborne Wind Energy Systems

In Ground-Generator Airborne Wind Energy Systems (GG-AWES) electrical energy is produced exploiting aerodynamic forces that are transmitted from the aircraft to the ground through ropes. As previously anticipated, GG-AWESs can be distinguished in devices with fixed or moving-ground-station.

Fixed-ground-station GG-AWES (or Pumping Kite Generators) are among the most exhaustively studied by private companies and academic research laboratories. Energy conversion is achieved with a two-phase cycle composed by a generation phase, in which electrical energy is produced, and a recovery phase, in which a smaller amount of energy is consumed. In these systems, the ropes, which are subjected to traction forces, are wound on winches that, in turn, are connected to motor-generators axes. During the generation phase, the aircraft is driven in a way to produce a lift force and consequently a traction (unwinding) force on the ropes that induce the rotation of the electrical generators. For the generation phase, the most used mode of flight is the crosswind flight with circular or the so-called eight-shaped paths. As compared to a non-crosswind flight (with the aircraft in a static angular position in the sky), this mode induces a stronger apparent wind on the aircraft that increases the pulling force acting on the rope. In the recovery phase motors rewind the ropes bringing the aircraft back to its original position from the ground. In order to have a positive balance, the net energy produced in the generation phase has to be larger than the energy spent in the recovery phase. This is guaranteed by a control system that adjusts the aerodynamic characteristics of the aircraft and controls its flight path in a way to maximize the energy produced in the generation phase and to minimize the energy consumed in the recovery phase.

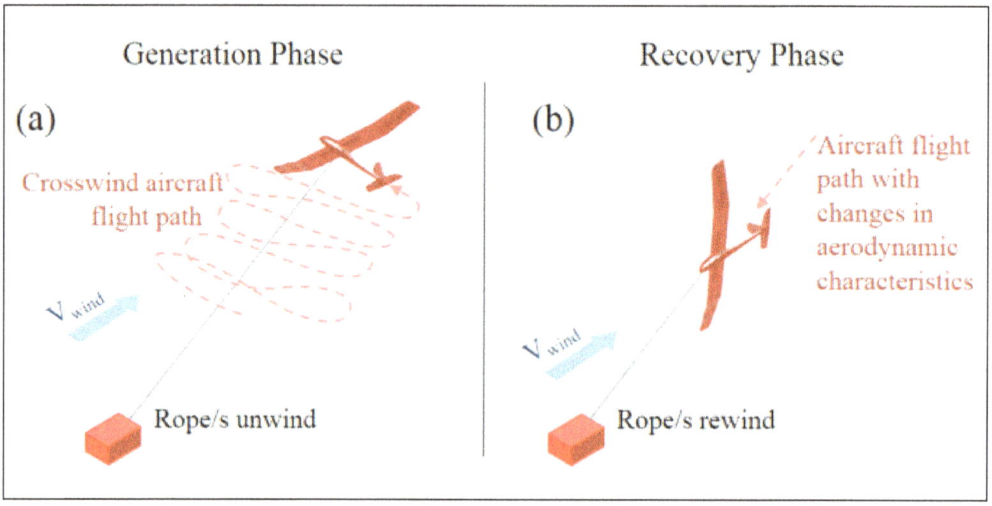

Scheme of the two-phase discontinuous energy production for GG AWESs. (a) The energy generation phase occurs during the unwinding of the ropes as the aircraft performs a crosswind flight. (b) The recovery phase is performed in order to minimize the energy consumed for the recovery.

Pumping kite generators present a highly discontinuous power output, with long alternating time-periods (in the order of tens of seconds) of energy generation and consumption. Such an unattractive feature makes it necessary to resort to electrical rectification means like batteries or large capacitors. The deployment of multiple AWES in large high-altitude wind energy farms could significantly reduce the size of electrical storage needed.

Moving-ground-station GG-AWES are generally more complex systems that aim at providing an always positive power flow which makes it possible to simplify their connection to the grid. There are different concepts of moving-ground-station GG-AWESs but no working prototype has been developed up to date and only one prototype is currently under development. Differently from the pumping generator, for moving-ground-station systems, the rope winding and unwinding is not producing/consuming significant power but is eventually used only to control the aircraft trajectory. The generation takes place thanks to the traction force of ropes that induces the rotation (or linear motion) of a generator that exploits the ground station movement rather than the rope winding mechanism.

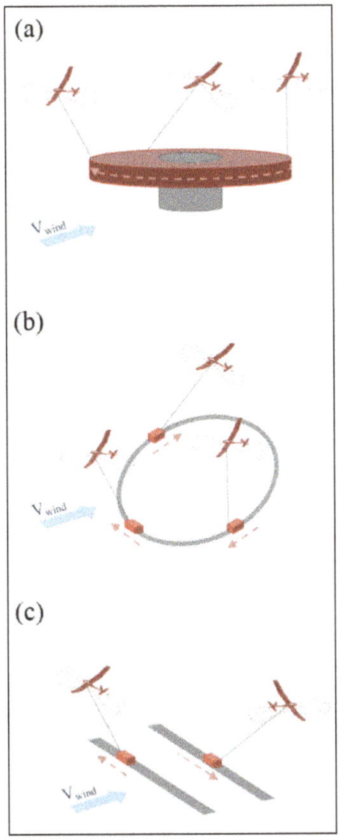

Scheme of three different concepts of moving-ground-station GG-AWES. (a) Vertical axis generator: ground stations are fixed on the periphery of the rotor of a vertical axis generator. (b) Closed loop rail: ground stations are fixed on trolleys that move along a closed loop rail. (c) Open loop rail: ground stations are fixed on trolleys that move along a open loop rail.

Basically, there are two kinds of moving-ground-station GG-AWES:

- 'Vertical axis generator': Where ground stations are fixed on the periphery of the rotor of a large electric generator with vertical axis. In this case, the aircraft forces make the ground stations rotate together with the rotor, which in turn transmits torque to the generator.
- 'Rail generators' (closed loop rail or open loop rail): Where ground stations are integrated on rail vehicles and electric energy is generated from vehicle motion. In these systems, energy generation looks like a reverse operation of an electric train.

Ground-Gen Systems Architectures and Aircraft

In GG systems the aircraft transmits mechanical power to the ground by converting wind aerodynamic forces into rope tensile forces. The different concepts that were prototyped are listed in figure below; examples of aircraft of GG systems that are currently under development are presented in figure. They exploit aerodynamic lift forces generated by the wind on their surfaces/wings.

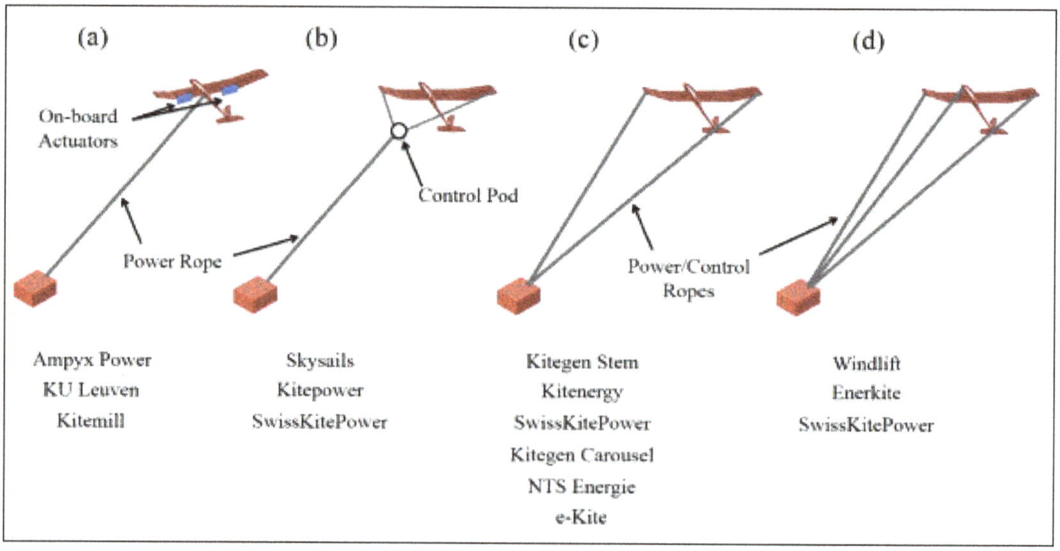

Control layout of crosswind GG-AWESs. (a) With on-board control actuators; (b) with flying control pod; (c) control through power ropes; (d) with additional control rope.

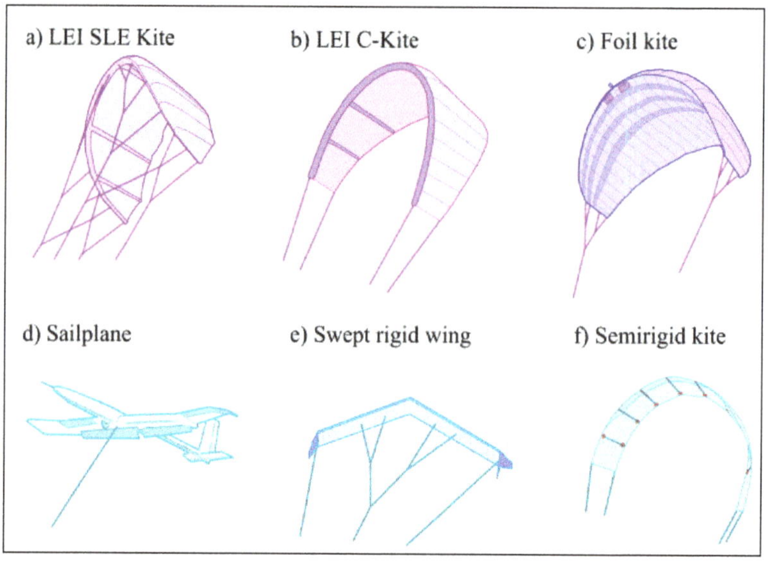

Different types of aircraft in Ground-Gen systems: (a) LEI SLE (Leading Edge Inflatable, Supported Leading Edge) Kite; (b) LEI C-kite; (c) Foil Kite, design from Skysails; (d) Glider, design from Ampyx Power; (e) Swept rigid wing, design from Enerkite; (f) Semi-rigid wing, design from Kitegen.

The aircraft is connected to the ground by at least one power-rope that is responsible for transmitting the lift force (and the harvested power) to the ground station. The flight trajectory can be controlled by means of on-board actuators, or with a control pod, or by regulating the tension of the same power-ropes, or with thinner control-ropes.

There are also two GG concepts that are worth mentioning: one uses parachutes which exploit aerodynamics drag forces, the other uses rotating aerostats which exploit the Magnus effect.

The most important aircraft used for GG systems are here listed:

- Leading Edge Inflatable (LEI) kites: Are single layer kites whose flexural stiffness is enhanced by inflatable structures on the leading edge. Mainly two kinds of LEI kites are used in AWESs:
    - Supported Leading Edge (SLE) kites: Are LEI kites with at least one bridle which supports the leading edge close to its central part. In comparison with C-kites (that are described in the following), the traction force of the central bridles makes the wing flat in its central region and this is claimed to increase the wing aerodynamic efficiency.
    - C-kites: Which are generally controlled by four main bridles directly attached to extreme lateral points of the kite edges. In pumping generators, the C-kite is held with either one, two or three ropes. In generators with one rope, the rope is connected to both the leading edge bridles, while trailing edge bridles are controlled by a 'control pod' (i.e. a flying box with one or more actuators) attached to the rope a few meters below the kite. The micro-winches inside the control pod are used to steer the kite and control the angle of attack. In case of two ropes, left bridles converge in one rope and right bridles converge in the second rope. The angle of incidence is fixed and the kite steers due to the difference in the ropes tension. In case of three ropes, there is one rope for each trailing edge bridle and one rope connected to the leading edge bridles. In this case, kite steering and angle of attack can be controlled from the ground.

    The stiffened tube-like structure of LEI kites is especially useful for take-off and landing maneuvers when the wing is not yet supported by wind pressure. The ease of handling is very appreciated also during small-scale prototyping and subsystem testing. However LEI kites have severe scalability issues as the tube diameter needs to be oversized in case of large wings.

- Foil kites (also called ram-air kites): Are derived from parafoils. These double-layer kites are made of canopy cells which run from the leading edge to the trailing edge. Cells (some or all) are open on the leading edge in a way that the air inflates all cells during the flight and gives the kite the necessary stiffness. Bridles are grouped in different lines, frequently three: one central and two laterals. With respect to LEI kites, foil wings have a better aerodynamic efficiency despite the higher number of bridles and can be one order of magnitude larger in size.

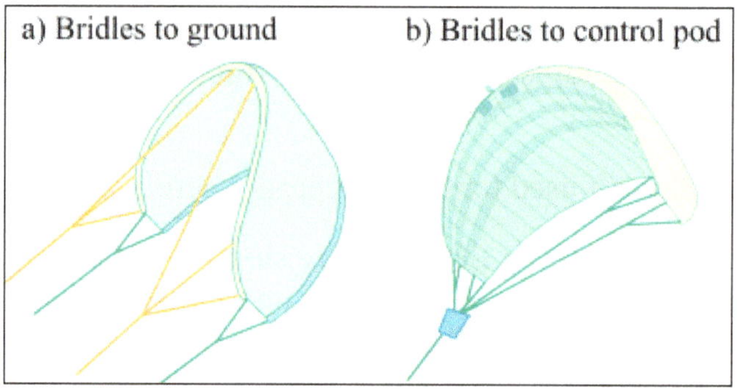

Control of bridles tension. (a) Control bridles are attached to the leading and trailing edges of a LEI SLE kite. (b) A control pod can be used to control the flight trajectory and angle of attack.

- Delta kites: Are similar to hang glider wings. They are made by a single layer of fabric material reinforced by a rigid frame. Compared with LEI or foil kites, this kind of aircraft has a better aerodynamic efficiency which in turn results in a higher efficiency of wind power extraction. On the other hand, their rigid frame has to resist to mechanical bending stresses which, in case of high aerodynamic forces, make it necessary to use thick and strong spars which increase the aircraft weight, cost and minimum take-off wind speed.

  Durability for fabric wings such as LEI, foil and delta kites, is an issue. Performance is compromised soon and lifetime is usually around several hundred hours.

- Gliders: Can also be used as GG aircraft. Like delta kites, their wings are subject to bending moment during the tethered flight. Gliders, and more generally rigid wings, have excellent aerodynamic performance, although they are heavier and more expensive. Lifetime with regular maintenance is several decades.

- Swept rigid wings: Are gliders without fuselage and tail control surfaces. Flight stability is most likely achieved thanks to the bridle system and the sweep angle.

- Semi-rigid wings: Are also under investigation by the Italian company Kitegen Research. They are composed of multiple short rigid modules that are hinged to each other. The resulting structure is lighter than straight rigid wings and more aerodynamically efficient and durable than fabric kites.

- Special design kites: Kiteplanes and Tensairity Kites are projects developed by TUDelft (The Netherlands) and EMPA (Research Center for Synergetic Structures, ETH Zurich), that aim at increasing the aerodynamic efficiency of arch kites without using rigid spars.

Fixed-ground-station Systems under Development

Table: This subsection provides a list of fixed-ground-station GG AWES which are summarized in table.

Ground-Gem HAWECs								
Fixed/Moving		Airborne system	Company	Power Class	Main Force	Actuators		N. Ropes
						Ca ground	Airborne	
Fixed ground station – Pumping System		Leading edge inflatable kite	KiteGen Stem	kW	Lift	X		2
			WindLift	kW	Lift	X		3
			Kitenergy	kW	Lift	X		2
			Swiss Kite Power 2	kW	Lift	X		3
			Kite Power	kW	Lift		X	1
			Swiss Kite Power 1	kW	Lift		X	1
		Foil Kite	Sky Sails Power	kW-MW	Lift		X	1
			Enerkite	kW	Lift	X		3
		Delta Kite	Enerkite	kW	Lift	X		3
		Swept Rigid Wing	Enerkite	kW	Lift	X		3
		Glider	Ampyx power	Kw-MW	Lift		X	1
			e-Kite	kW	Lift	X		2
			Kitemill	kW	Lift		X	1
		Glider with rotors	Tewing Tec	kW	Lift		X	2
		Semi-rigid Wing	KiteGen Stem	MW	Lift	X		2
		Parachute	Guangdong Tech	MW	Drag		X	2
		Aerostat	Omnidea	kW	Megnus Effect		X	2
Moving ground station	Rail	Leading edge Inflatable kite	KiteGen Rail Carousel	MW-GW				2
			Kitenergy	MW	Lift	X		2
		Foil kite	NTS		Lift	X		4
	Axial	Leading edge inflatable kite	KiteGen Carousel	MW	Lift	X		2

KiteGen Research

The Italian KiteGen Research (KGR) was one of the first companies to test a prototype of Ground-Gen AWES. KGR technology is based on a C-Kite integrating on board electronics with sensor and is controlled by two power-ropes from a control station on the ground. The first prototype, named KSU1 (acronym for Kite Steering Unit), was successfully demonstrated in 2006. After a few years of tests, the company focused on the development of a new generator, named 'KiteGen Stem', with a nominal power of 3 MW. In this system, the ropes are wound on special winches and are driven by a pulley system through a 20 m flexible rod, called 'stem', to an arch-kite or a semi-rigid wing. The stem is linked to the top of the control station through a pivot joint with horizontal axis. The most important functions of the stem are: (1) supporting and holding the kite and (2) damping peak forces in the rope that arise during wind-gusts. The entire control station can make azimuthal rotations so the stem has two degrees of freedom relative to the ground. The 'Stem' concept was first patented in 2008 and is now used by more and more companies and universities.

At the beginning of the take-off maneuvers, the kite is hanged upside down at the end of the stem. Once the kite has taken off, the production phase starts: the automatic control drives the kite acting on the two ropes, the kite makes a crosswind flight with 'eight shape' paths; at the same time ropes are unwound causing the winches to rotate; the motor-generators transform mechanical power into electric power. The company aims at retracting the cables with minimum energy consumption thanks to a special maneuver called 'side-slip' or 'flagging'. Side-slip is a different flight mode where the kite aerodynamic lift force is cleared by rewinding at first one rope before the other, which makes the kite lose lift and 'stall' and then, once fully stalled, both ropes are rewound at the same speed and the kite precipitates flying sideways. This maneuver can be done with flexible foil kites or semi-rigid wings. In this phase, the power absorbed by motor-generators is given by rope rewind speed multiplied by the resulting aerodynamic drag force of the side-slip flying mode. This power consumption would be a small percentage of the power produced in the production phase. After rewinding a certain length of the ropes (less than the total rope length in order to exploit only the highest winds) another special maneuver restores the gliding flight and the aerodynamic lift force on the kite. At this point one pumping cycle ends and a new production phase starts.

KGR patented and is developing special aerodynamic ropes in order to increase their endurance and to increase system performances. KGR also plans to use the Kitegen Stem technology to produce an offshore AWES since offshore AWESs are very promising.

Kitenergy

Kitenergy, was founded by a former KiteGen partner and is also developing a similar concept by controlling a foil kite with two ropes. The prototype of the company features 60 kW of rated power. Kitenergy submitted also a different GG-AWES patent that consists in a system based on a single motor-generator which controls winding and unwinding of two (or more) cables and another actuator that introduces a differential control action of the employed cables. Another prototype developed by its co-founder, Lorenzo Fagiano, achieved 4 h of consecutive autonomous flight with no power production at University of California at Santa Barbara in 2012.

SkySails Power

The German company SkySails GmbH is developing a wind propulsion system for cargo vessels based on kites. A few years ago a new division of the company 'SkySails Power' has been created to develop Ground-Gen AWES based on the technology used in SkySails vessel propulsion system. Two products are under development: a mobile AWES having a capacity between 250 kW and 1 MW, and an offshore AWES with a capacity from 1 to 3.5 MW. SkySails' AWES is based on a foil kite controlled with one rope and a control pod which controls the lengths of kite bridles for steering the kite and changing its angle of attack. Control pod power and communication with the ground station is provided via electric cables embedded in the rope. SkySails also has a patented launch and recovery systemdesigned for packing the kite in a storage compartment. It is composed by a telescopic mast with a special device on its top that is able to grab, keep and release the central point of the kite leading edge. When the system is off, the mast is compacted in the storage compartment with the kite deflated. At the beginning of the launching operation, the mast extends out vertically bringing the deflated kite some meters above the ground (or the sea level). The kite is then inflated to have appropriate shape and stiffness for the production phase. Kite take-off exploits only the natural wind lift force on the kite: the system at the top of the mast releases the kite leading edge, the pod starts to control the flight and the winch releases the rope letting the kite reach the operating altitude. While the energy production phase is similar to that of the KGR generator, SkySails has a different recovery phase. Specifically, SkySails uses high speed winching during reel-in while the kite is kept at the edge of the wind window. The kite is then winched directly against the wind without changing the kite angle of attack. Though it might seem counter-intuitive at first, this kind of recovery phase has proven to be competitive.

TwingTec

The Swiss company Twingtec is developing a 100 kW GG-AWES. After having tried several concepts including soft wings and rigid wings, the team is now tackling the problem of automating take-off and landing with an innovative concept: a glider with embedded rotors having rotational axis perpendicular to the wing plane. The rotors are used during take-off and landing. The company plans to have the generator and power conversion hardware inside a standard 20-foot shipping container in order to easily target off-grid and remote markets. The AWES will supply continuous and reliable electrical power thanks to the integration with conventional diesel generators.

TU Delft

At Delft University of Technology, the first research in Airborne Wind Energy was started by the former astronaut, Professor Ockels, in 1996. A dedicated research group was initiated by Ockels in 2004 with the aim to advance the technology to the prototype stage.

Recently, Delft University of Technology and Karlsruhe University of Applied Sciences have initiated a joint project to continue the development and testing of a mobile 20 kW experimental pumping kite generator. A main objective of this project is to improve the reliability and robustness of the technology and to demonstrate in the next months a continuous operation of 24 h. At present, they use the third version of a special design LEI kite, co-developed with Genetrix/Martial Camblong, of 25 m² wing surface area. Together with an automatic launch setup, the wing demonstrated fully automatic operation of their 20 kW system in 2012. Like SkySails' system, this prototype is based on a single tether and an airborne control pod but they also control the angle of attack for powering and depowering the wing during production and recovery phase, respectively. An automatic launch and retrieval system for 100 m² LEI kites is under development.

In the past, the research group tested several kinds of wings such as foil kites and kiteplanes. TU Delft also tested an alternative device for controlling the kite: a cart-and-rail system attached to the tips of a ram-air wing and used to shift the attachment point of the two bridle lines. By that system, the wing could be steered and depowered with a minimal investment of energy. Ultimately, the concept was too complex and too sensitive to deviations from nominal operation.

Ampyx Power

The first company that developed a pumping glider generator is the Dutch Ampyx Power. After several prototypes, they are currently developing and testing two 5.5 m 'PowerPlanes' the AP-2A1 and the AP-2A2. They are two officially registered aircraft that are automatically controlled with state of the art avionics. They are constructed with a carbon fiber body and a carbon backbone truss which houses onboard electronics with sensors and actuators. Onboard actuators can drive a rudder, an elevator and four flaperons. One rope connects the glider to a single winch in the ground station. Ampyx Power is actually one of the few companies which has already developed an AWES that is able to automatically perform the sequence of glider take-off, pumping cycles and landing. Take-off maneuver sees the glider lying on the ground facing the ground station at some meters of distance. As the winch starts exerting traction force on the rope, the glider moves on the ground and, as soon as the lift forces exceed the weight forces, the glider takes off. They also installed a catapult for take-off and they have a propulsion system to climb up. The glider flight is fully autonomous during normal operations even though, for safety reasons, it can be occasionally controlled wirelessly from the ground thanks to a backup autopilot. The pumping cycles are similar to those of a kite. Glider landing is similar to that of an airplane and is being equipped with an arresting line so as to stop the glider in a right position for a new take-off. During a test campaign in November 2012, the system demonstrated an average power production of 6 kW with peaks of over 15 kW (earlier tests showed peak in power production of 30 kW). Ampyx has started the design of its first commercial product: a 35 m wingspan AP-4 PowerPlane with a 'wind turbine equivalent' power of 2 MW.

EnerKite

The German company EnerKite developed a portable pumping kite generator with rated continuous power of 30 kW. The ground station is installed on a truck through a pivotal joint which allows azimuthal rotations. EnerKite demonstrator uses mainly a foil kite, but a delta kite and a swept rigid wing are also under investigation and testing. The aircraft does not have on-board sensors and is controlled from the ground with three ropes according to the scheme of figure. EnerKite is now developing an autonomous launch and landing system for semi-rigid wings. The company plans to produce a 100 kW and a 500 kW system.

Windlift

The US Company Windlift has a concept similar to that of Enerkite. Their 12 kW prototype uses SLE kites. They aim to sell their product to the military and to off-grid locations.

New Companies

The AWE community is constantly growing and for every company that goes out of business there are a few that are born. Here are some startup companies that are worth mentioning.

e-Kite was founded in 2013 in the Netherlands and developed a 50 kW GG-AWES based on a direct drive generator. The company is now building a 2-ropes rigid wing that will fly at low altitude.

Enevate is a Dutch 4-people startup that is mainly focused on bringing the TU Delft GG-AWES to the next step towards a commercial product.

Kitemill, Norway, started the development of a GG-AWES. The company switched early on to a 1-cable rigid wing system with on-board actuators after having faced controllability and durability issues with soft materials.

eWind solutions is a US company that is developing an unconventional, low altitude, rigid wing GG-AWES.

KU Leuven

KU Leuven has been actively doing research in AWESs since 2006. After significant theoretical contributions, the team developed a test bench to launch a tethered glider with a novel procedure. Before take-off, the glider is held at the end of a rotating arm. When the arm starts rotating, the glider is brought to flying speed and the tether is released allowing the glider to gain altitude. They are currently developing a larger experimental test set-up, 2 m long with a 10 kW winch.

SwissKitePower

SwissKitePower was a collaborative research and development project started in Switzerland in 2009. It involved four laboratories of different Swiss universities: FHNW, EMPA (Swiss Federal Laboratories for Materials Science and Technology), ETH (Federal Politechnique of Zurich) and EPFL (École Polytechnique Fédérale de Lausanne). The first prototypes, tested between 2009 and 2011, were based on a C-kite controlled by one rope and a control pod. The initial system worked according to the scheme of figure., similarly to KitePower and SkySails prototypes. In 2012, SwissKitePower developed a new ground station with three winches that can be used to test kites with 1, 2 or 3 lines. They also tested SLE kites and tensairity kites. The project ended in 2013 and since then FHNW is working in collaboration with the company TwingTec.

NASA Langley

At Langley Research Center, the US space agency NASA conducted a study about wind energy harvesting from airborne platforms after which they developed an AWES demonstrator based on a kite controlled by two ropes and having a vision-based system and sensors located on the ground.

Others

In addition to the main prototypes listed above, there are several other systems that have been built:

- Wind tunnel tests of small scale non-crosswind generation, and outdoor crosswind generation tests with a SLE kite, of GIPSA-lab/CNRS, University of Grenoble.
- Kite control project of CCNR at Sussex University, UK.
- EHAWK (Electricity from High Altitude Wind with Kites) project of Department of Mechanical Engineering of Rowan University.
- Kite powered water pump of Worcester Polytechnic Institute.

Moving-ground-station Systems under Development

In addition to pumping systems, a number of AWES concepts with moving-ground-station have been proposed. Their main advantage is the ability to produce energy continuously or nearly continuously. However, only a few companies are working on AWESs with moving-ground-station and there are more patents and studies than prototypes under development.

KiteGen Research

The first moving-ground-station architecture which is based on a vertical axis generator has been proposed back in 2004 by Sequoia Automation and acquired by KGR. This AWES concept is based on the architecture described in figure. During operations, lift forces are transmitted to a rotating frame inducing a torque around the main vertical axis. Torque and rotation are converted into electricity by the electric generator. This system can be seen as a vertical axis wind turbine driven by forces which come from tethered aircraft. There is no prototype under development, but the concept has been studied in a simulation showing that 100 kites with 500 m^2 area could generate 1000 MW of average power in a wind with speed of 12 m/s. The considered generator would have a 1500 m radius, occupying a territory about 50 times smaller and costing about 30 times less than a farm of wind turbines with the same nominal power.

NTS Energie

An alternative system based on ground stations that moves on closed track circuits is proposed by KGR and by the German company NTS Energie und Transportsysteme. Starting from September 2011, NTS tested a prototype where 4-rope kites are controlled by a vehicle which moves on a 400 m flat-bed straight railway track. They are able to produce up to 1 kW per m^2 of wing area and they tested kites up to 40 m^2. The final product should have a closed loop railway where more vehicles run independently.

Kitenergy

Another rail concept is proposed by Kitenergy and it is based on ideas published in 2004 in Drachen Foundation journal. The concept is based on a straight linear rail fixed on the ground with a pivotal joint. The rail direction is then adjusted perpendicular to the main direction of the wind. The ground station of the system is mounted on a wheeled vehicle which moves along the straight rail, under the kite traction forces, back and forth from one side to the other. The power is extracted from electromagnetic rotational generators on the wheels of the vehicle or from linear electromagnetic generators on the rail. The power production is not fully continuous because during the inversion of vehicle direction the power production will not only decrease to zero, but it could also be slightly negative. Nevertheless the kite inversion maneuver could be theoretically performed without the need of power consumption.

Laddermill

Although it cannot be considered a moving-ground-station device, it is important to mention that the first concept of continuous energy production AWES was the Laddermill concept envisaged by the former astronaut, Professor Ockels in 1996.

Fly-Gen Airborne Wind Energy Systems

In Fly-Gen AWESs, electric energy is produced onboard of the aircraft during its flight and it is transmitted to the ground trough one special rope which integrates electric cables. Electrical energy conversion in FG-AWESs is achieved using one or more specially designed wind turbines.

Aircraft in Fly-Gen Systems

Besides the general classification between crosswind and non-crosswind mode proposed in figure below. FG-AWESs can also be distinguished basing on their flying principles that are:

- Wings lift: Achieved with a tethered flight of special gliders or frames with multiple wings.

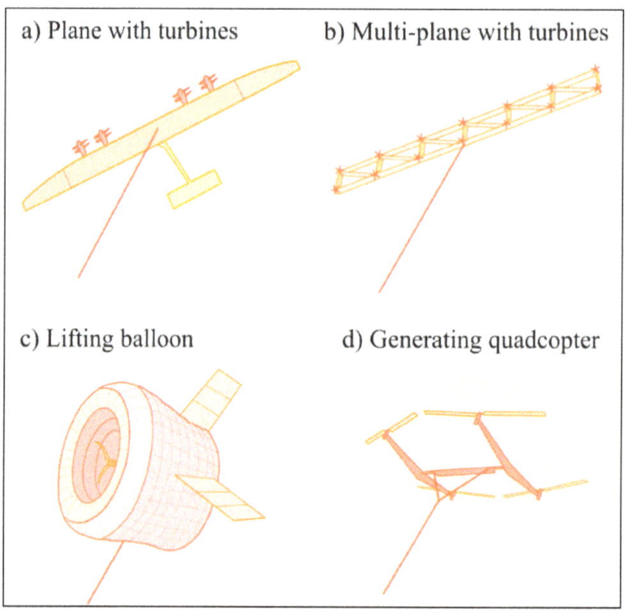

Different types of aircraft in Fly-Gen systems: (a) Plane with four turbines, design by Makani Power. (b) Aircraft composed by a frame of wings and turbines, design by Joby Energy. (c) Toroidal lifting aerostat with a wind turbine in the center, design by Altaeros Energies. (d) Static suspension quadrotor in autorotation, design by Sky WindPower.

- Buoyancy and static lift: Achieved with aerodynamically shaped aerostats filled with lighter-than-air gas.
- Rotor thrust: Achieved with the same turbines used for electrical power generation .

Aircraft in figure, fly crosswind and harvest the relative wind, while those in figure and fly non-crosswind and harvest the absolute wind.

There is also one FG concept that aims at exploiting high altitude wind energy not by using aerodynamic lift. It uses instead a rotating aerostat which exploits the Magnus effect.

Fly-Gen Systems under Development

This subsection provides a list of FG AWES which are summarized in figure.

Fly-Gen HAWECs				
General system description	Company	Flying principle	Type	Energy generation system
Turbines on a tethered aircraft	Makani Power	Wings lift	Crosswind	6/8 turbines
	Joby Energy	Wings lift	Crosswind	Several turbines
Tethered quadcopter	Sky Windpower	Rotors thrust	Non-crosswind	4 turbines
Turbine on a lighter than the air balloon	Altaeros Energies	Buoyancy	Non-crosswind	1 turbine
Magnus effect turbine	Omnidea	Buoyancy	Non-crosswind	Buoyant wind turbine

Fly-Gen AWESs. Summary of FG-AWESs.

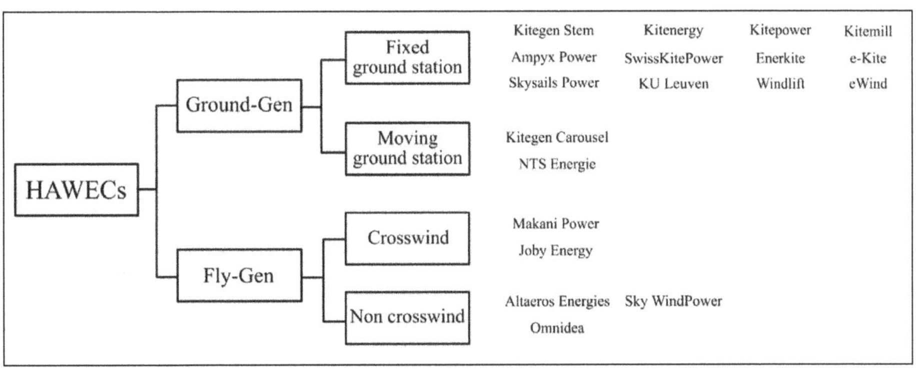

Classification of AWESs. The different AWESs concepts.

Loyd's First Mechanical Concept

One of the most famous and old idea of exploiting wind energy using turbines on a kite belongs to Loyd who calculated that wind turbines installed on a crosswind flying kite could be able to generate up to 5 times the power produced by equivalent turbines installed on the ground. He also patented his idea in 1978. Loyd's concept foresees a reciprocating wind driven apparatus, similar to a multi propeller plane, with a plurality of ropes linking the aircraft to a ground station.

Makani Power

After about twenty-five years from Loyd's work, Makani Power Inc. has started the development of its Airborne Wind Turbine (AWT) prototypes. In nine years, Makani tested several AWESs concepts including Ground-Gen, single rope, multiple rope, movable ground station on rails, soft wings and rigid wings. During these years, the company filed several patents where an electric and modern version of Loyd's idea has been enriched with a tether tension sensor, an aerodynamic cable, and with a new idea of a bimodal flight that has been invented to solve take-off and landing issues. In the bimodal flight the AWT takes off with the wing plane in a vertical position, driven by propellers thrust. This flight mode is similar to a quadcopter flight and rotors on AWT are used as engines. Once all the rope length has been unwound, the AWT changes flight mode becoming a tethered flight airplane. In this second flight mode a circular flight path is powered by the wind itself and rotors on AWT are used as generators to convert power from the wind. During this phase the cable length is fixed. In order to land, a new change of flight mode is performed, and the AWT lands as a quadcopter. Makani has developed and tested its 8 m, 20 kW demonstrator, called 'Wing 7' that showed the capability of fully automatic operations and power production. After these results, in early 2013 Makani was acquired by Google. Makani is currently developing a 600 kW prototype, 'the M600'. The M600 AWT has eight turbines, each with five propeller blades, and has a wingspan of 28 m. The prototype is now undergoing testing. After M600, Makani plans to produce an offshore commercial version of AWT with a nominal power of 5 MW featuring 6 turbines and a wingspan of 65 m.

Joby Energy

Founded in 2008, Joby Energy Inc. is another US company which is developing a FG-AWES. The main difference between Joby and Makani is that the tethered airborne vehicle is a multi-frame structure with embedded airfoils. Turbines are installed in the joints of the frame. In Joby's concept, the system could be adapted to be assembled with modular components, constructed from multiple similar frames with turbines. The power generation method and the take-off and landing maneuvers are similar to those of Makani concept. Joby also patented an aerodynamic rope for its system. In 2009 and 2010, Joby tested different small scale prototypes.

Altaeros Energies

Another project based on flying wind turbines in a stationary position has been developed by Altaeros Energies, a Massachusetts-based business led by MIT and Harvard alumni. In this case, instead of using wings lift to fly, they use a ring shaped aerostat with a wind turbine installed in its interior. The whole generator is lighter than the air, so the take-off and landing maneuvers are simplified, and the only remaining issue is the stabilization of the generator in the right position relative to the wind. The aerostat is aerodynamically shaped so that the absolute wind generates lift that helps keeping a high angle of altitude together with the buoyancy force. After their energy production tests in 2012, Altaeros is additionally working on multiple rotor generators with different lighter-than-air craft configurations.

Sky Windpower

Sky Windpower Inc. proposed a different kind of tethered craft called 'Flying Electric Generator' (FEG) which is similar to a large quadrotor with at least three identical rotors mounted on an airframe that is linked to a ground station with a rope having inner electrical cables. Their concept was the first AWES to be tested in 1986 at University of Sidney. Take-off and landing maneuvers are similar to those of Makani's and Joby's generators, but FEG operation as generator is different. Once it reaches the operational altitude, the frame is inclined at an adjustable controllable angle relative to the wind (up to 50 deg) and the rotors switch the functioning mode from motor to generator. At this inclined position, the rotors receive from their lower side a projection of the natural wind parallel to their axes. This projection of wind allows autorotation, thus generating both electricity and thrust. Electricity flows to and from the FEG through the cable. Sky Windpower tested two FEG prototypes. They claimed that a typical minimum wind speed for autorotation and energy generation is around 10 m/s at an operational altitude of 4600 m. Unfortunately the company went recently out of business.

Crosswind Flight: The Key to Large Scale Deployment

One of the most important reasons why AWESs are so attractive is their theoretical capability of achieving the megawatt scale with a single plant. For example in a 34 MW plant is envisaged with a tethered Airbus A380, and many other publications present theoretical analyses with MW scale AWES. This scalability feature is rare in renewable energies and is the key to successful commercial development.

Modelling the principle of crosswind flight is the first necessary step towards the understanding of AWESs and their potential. A well known basic model is explained for the case of Ground-Gen and Fly-Gen crosswind AWESs.

Only crosswind generation is analyzed because it was demonstrated that it can provide a power one or two orders of magnitude higher than non-crosswind generation. AWESs concepts that exploit crosswind power have therefore a strong competitive advantage over non-crosswind concepts in terms of available power and, therefore, in the economics of the whole system.

Advantages and Challenges of Wind Energy

Wind energy offers many advantages, which explains why it's one of the fastest-growing energy sources in the world. Research efforts are aimed at addressing the challenges to greater use of wind energy.

Advantages of Wind Power

- Wind power is cost-effective: Land-based utility-scale wind is one of the lowest-priced energy sources available today, costing between two and six cents per kilowatt-hour, depending on the wind resource and the particular project's financing. Because the electricity from wind farms is sold at a fixed price over a long period of time (e.g. 20+ years) and its fuel is free, wind energy mitigates the price uncertainty that fuel costs add to traditional sources of energy.
- Wind creates jobs: The U.S. wind sector employed more than 100,000 workers in 2016, and wind turbine technician is one of the fastest-growing American jobs of the decade. wind has the potential to support more than 600,000 jobs in manufacturing, installation, maintenance, and supporting services by 2050.
- Wind enables U.S. industry growth and U.S. competitiveness: Wind has an annual economic impact of about $20 billion on the U.S. economy, The United States has a vast domestic resources and a highly-skilled workforce, and can compete globally in the clean energy economy.
- It's a clean fuel source: Wind energy doesn't pollute the air like power plants that rely on combustion of fossil fuels, such as coal or natural gas, which emit particulate matter, nitrogen oxides, and sulfur dioxide—causing human health problems and economic damages. Wind turbines don't produce atmospheric emissions that cause acid rain, smog, or greenhouse gases.
- Wind is a domestic source of energy: The nation's wind supply is abundant and inexhaustible. Over the past 10 years, cumulative wind power capacity in the United States increased an average of 30% per year, and wind now has the largest renewable generation capacity of all renewables in the United States.
- It's sustainable: Wind is actually a form of solar energy. Winds are caused by the heating of the atmosphere by the sun, the rotation of the Earth, and the Earth's surface irregularities. For as long as the sun shines and the wind blows, the energy produced can be harnessed to send power across the grid.

- Wind turbines can be built on existing farms or ranches: This greatly benefits the economy in rural areas, where most of the best wind sites are found. Farmers and ranchers can continue to work the land because the wind turbines use only a fraction of the land. Wind power plant owners make rent payments to the farmer or rancher for the use of the land, providing landowners with additional income.

Challenges of Wind Power

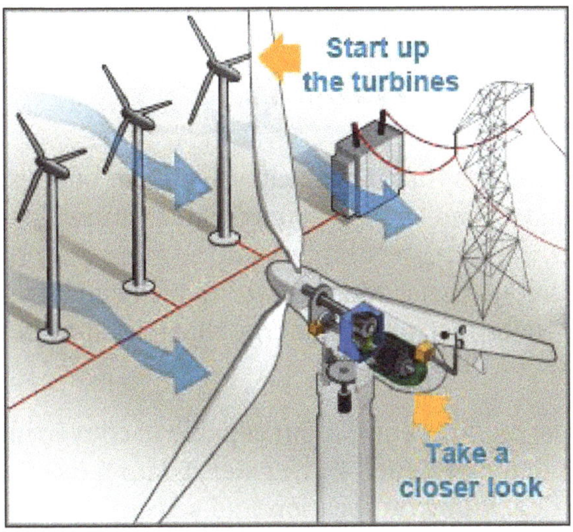

- Wind power must still compete with conventional generation sources on a cost basis: Depending on how energetic a wind site is, the wind farm might not be cost competitive. Even though the cost of wind power has decreased dramatically in the past 10 years, the technology requires a higher initial investment than fossil-fueled generators.
- Good wind sites are often located in remote locations, far from cities where the electricity is needed: Transmission lines must be built to bring the electricity from the wind farm to the city. However, building just a few already-proposed transmission lines could significantly reduce the costs of expanding wind energy.
- Wind resource development might not be the most profitable use of the land: Land suitable for wind-turbine installation must compete with alternative uses for the land, which might be more highly valued than electricity generation.
- Turbines might cause noise and aesthetic pollution: Although wind power plants have relatively little impact on the environment compared to conventional power plants, concern exists over the noise produced by the turbine blades and visual impacts to the landscape.

- Turbine blades could damage local wildlife: Birds have been killed by flying into spinning turbine blades. Most of these problems have been resolved or greatly reduced through technological development or by properly siting wind plants.

This aerial view of a wind turbine plant shows how a group of wind turbines can make electricity for the utility grid. The electricity is sent through transmission and distribution lines to homes, businesses, schools, and so on.

Ways to Conserve Wind Energy

One of the benefits of wind is that it's a renewable form of energy, meaning it can be used over and over again a wind energy resource site. Nature also uses wind as a renewable energy source, as the Earth's rotations and varying atmospheric pressures keep air moving on a continuous basis. Wind power that's been harnessed can be converted into a form of kinetic or mechanical energy that's used to power different types of equipment.

Harnessing wind power requires equipment capable of converting its energy, and storing it inside a system that's designed to transfer it into a useable form as needed. As wind intensities vary depending on location and setting, open rural-type areas create the best conditions for harnessing and storing wind for future use.

Generators

One way to conserve wind energy is by using a generator system that's powered by wind turbines, A wind turbine is made up of a rotor, a series of blades and a shaft. As the wind blows against the blades, the rotor turns the shaft. When the shaft is connected to a generator, wind energy is converted into mechanical energy, which causes the generator to spin. As the generator spins, electrical power or energy is created.

Generator equipment is also capable of storing electrical power for later use, much like a battery cell stores energy. Open areas free of buildings and trees create the best conditions for a wind generator to work. Turbine systems vary in the amount of power they can generate. Turbine capacity is determined by the number and size of the rotor blades, as larger blades and a greater number of blades can generate more power.

Wind and Solar

The use of solar panel systems alongside wind generators is another means of conserving wind energy, according to Wind Job. Solar panel systems convert sunlight into useable forms using photovoltaic, or solar cell, technology. Solar cells contain a material that's capable of generating an electric current when exposed to sunlight. This current can then be used to power devices like a pump, a heating system or an electrical grid.

Solar and wind generator systems can be used interchangeably in cases where sunlight exposure may be limited. A solar wind system is no doubt more complicated to set up than a wind generator alone; however, the use of both energy sources can provide a more steady source of power.

References

- Wind-power: mpoweruk.com, Retrieved 31 March, 2019
- Windmills: renewableresourcescoalition.org, Retrieved 14 July, 2019
- Offshore-wind-power, electricity, businesses: engie.com, Retrieved 17 May, 2019
- What-are-the-pros-and-cons-of-onshore-wind-energy, faqs, granthaminstitute: lse.ac.uk, Retrieved 19 April, 2019
- Advantages-and-challenges-wind-energy, wind: energy.gov, Retrieved 5 February, 2019
- Ways-to-conserve-wind-energy: sciencing.com, Retrieved 26 July, 2019

Chapter 2

Wind Turbine and its Types

Wind turbines are the devices used for the conversion of kinetic energy of wind into electrical energy. The two main types of wind turbines are horizontal-axis turbines and vertical-axis turbines. The topics elaborated in this chapter will help in gaining a better perspective about these types of wind turbines.

Wind turbines are the evolution of the classic windmills that can be seen in more rural areas of the world. Their purpose is to reduce reliance on fossil fuels to create energy and also to create energy in a less wasteful manner. They operate by using the kinetic energy of the wind, which pushes the blades of the turbine and spins a motor that converts the kinetic energy into electrical energy for consumer use.

Wind Turbines are rotating machines that can be used directly for grinding or can be used to generate electricity from the kinetic power of the wind. They provide the clean and renewable energy for us of both home and office. Wind Turbines are a great way to save money and make the environment clean and green.

This process has been adapted for use for various applications and can be seen in use by boats, traffic signs, or whole communities that use a wind farm for power. The development of wind turbines is a major step towards overhauling the way we produce our energy.

Basically there are two types of wind generators, those with vertical axis and those with horizontal axis. They can be used to generate electricity both onshore and offshore.

Wind Turbines can be combined to form clusters called “wind farms” which are used by large companies to use that power as their backup. Apart from generating electricity they can also be used for grain-grinding, water pumping, charging batteries.

Historically, wind turbines were used for sailing, irrigation and grinding-grains. It was in the early 20th century that it was used for generating power. Today, large wind turbines can be seen in the rural areas or near the sea coast where speed of the wind is generally throughout the day. Device called wind resource assessment is used for estimating the wind speed.

Wind Turbine Components

Wind turbine systems are made up of many different pieces of equipment that all serve a purpose in delivering electricity where it is intended to go. Of those many different pieces, the below list serves as a general blueprint for the main components that can and often times will be found in wind turbine systems regardless of the type of design.

- Rotor: The rotor is made of blades that are attached to a center piece. The blades are shaped such that when the wind pushes against them they turn.
- Pitch Drive: Used to rotate blades to accommodate for high-speed wind.
- Nacelle: The rotor is attached to a housing unit called a nacelle, which protects various other components necessary to the wind turbine operation.
- Brake: Necessary to slow the rotor down.
- Low-Speed Shaft: Attaches to the rotor and turns as the rotor turns on a 1:1 ratio.
- Gear Box: Serves the same function as a car, the rotor spins slowly as the wind pushes against it and the gearbox or transmission increases that rotational speed for the generator.
- High-Speed Shaft: Attaches to the gearbox and generator and spins at a higher speed than the rotor or low-speed shaft.
- Generator: Actual mechanism that converts the rotational kinetic energy into electricity.
- Wind Vane: Detects direction of wind and adjusts the rotor and nacelle to compensate.
- Yaw Drive: Keeps the rotor and therefore the turbines facing the wind.
- Tower: Elevates the aforementioned components to an altitude that optimizes wind exposure.

How do Wind Turbines Work?

Wind Turbines operate on a system comprised of many critical components that allow kinetic wind energy to be transformed into electrical energy. No matter the type of wind turbine system, they all work off the same principle that allows a generator to produce electricity. This principle is that if magnets are rotated around a coil of wire, or a coil of wire rotates within a magnetic field fast enough electricity is produced.

The total contraption of magnets and conductors make up the generator. Using the wind to turn blades creates the force needed to turn the magnets or the coil of conductor which in turn creates electricity. Below is a step by step process that highlights the methods by which a wind turbine actually comes to produce electricity:

- Tower is constructed that puts the wind turbine system at the correct altitude where wind travels at a higher and more constant rate.
- Rotor blades are exposed to wind which forces them to start turning.
- As the rotor spins, the low-speed shaft, which is connected to a gearbox, spins at the same rate.
- The gearbox takes this slow rotational speed and through correct gearing turns it into a faster rotational speed.
- The high-speed shaft, which is on the outgoing end of the gearbox and connected to a generator, spins at a higher rate of speed.
- The generator spins at this high rate of speed which spins magnets around a coil of metal wire and generates electricity.
- The electricity travels from the generator through wires to the necessary applications whether it be direct appliances or a battery.

Types of Wind Turbines

There are two main types of wind turbines that can be seen in design and implementation in the wind energy industry today. The first and most common type is the horizontal axis wind turbine that relies on a horizontal shaft that runs perpendicular to the blades which spin vertically. These wind turbine systems can be seen in use in major wind farms as well as solo operations.

The second type which is less common among the wind energy industry is the vertical axis wind turbine. As one might be able to infer, the vertical axis turbine has a vertical shaft in which the blades or rotor are connected to and spin horizontally. There are many variations of the vertical axis wind turbine but the major benefit is that maintenance is easier because the gearbox and generator are more accessible.

Wind Turbine Generator

Wind Turbine Generator or WTG's is the electrical machine used to generate the electricity.

A low rpm electrical generator is used for converting the mechanical rotational power produced by the winds energy into usable electricity to supply our homes and is at the heart of any wind power system.

The conversion of the rotational mechanical power generated by the rotor blades (known as the prime mover) into useful electrical power for use in domestic power and lighting applications or to charge batteries can be accomplished by any one of the following major types of rotational electrical machines commonly used in a wind power generating systems:

- The direct current (DC) machine, also known as a Dynamo.
- The alternating current (AC) synchronous machine, also known as an AC Generator.
- The alternating current (AC) induction machine, also known as an Alternator.

All these electrical machines are electromechanical devices that work on Faraday's law of electromagnetic induction. That is they operate through the interaction of a magnetic flux and an electric current, or flow of charge. As this process is reversible, the same machine can be used as a conventional electrical motor for converting the electrical power into mechanical power, or as a generator converting the mechanical power back into the electrical power.

Wind Turbine Induction Generator.

The electrical machine most commonly used for wind turbines applications are those acting as generators, with synchronous generators and induction generators being commonly used in larger wind turbine generators, while smaller and home made wind turbines tend to use a low speed DC generator or Dynamo as they are small, cheap and a lot easier to connect up.

So does it make a difference what type of electrical generator we can use to produce wind power. The simple answer is both yes and no, as it all depends upon the type of system and application you want. The low voltage DC output from a generator or older style dynamo can be used to charge batteries while the higher AC sinusoidal output from an alternator can be connected directly to the local grid.

Also, the output voltage and power demand depends entirely upon the appliances you have and how you wish to use them. In addition, the location of the wind turbine generator, would the wind resource keep it constantly rotating for long periods of time or would the generator speed and therefore its output vary up and down with variations in the available wind.

Electricity Generation

A Wind Turbine Generator is what makes your electricity by converting mechanical energy into electrical energy. They do not create energy or produce more electrical energy than the amount of mechanical energy being used to spin the rotor blades. The greater the "load", or electrical demand placed on the generator, the more mechanical force is required to turn the rotor. This is why generators come in different sizes and produce differing amounts of electricity.

In the case of a "wind turbine generator", the wind pushes directly against the blades of the turbine, which converts the linear motion of the wind into the rotary motion necessary to spin the generators rotor and the harder the wind pushes, the more electrical energy can be generated. Then it is important to have a good wind turbine blade design to extract as much energy out of the wind as possible.

All electrical turbine generators work because of the effects of moving a magnetic field past an electrical coil. When electrons flow through an electrical coil, a magnetic field is created around it. Likewise, when a magnetic field moves past a coil of wire, a voltage is induced in the coil as defined by Faraday's law of magnetic induction causing electrons to flow.

Simple Generator using Magnetic Induction

Then we can see that by moving a magnet past a single loop of wire, a voltage known as and emf (electro-motive force) is induced within the wire loop due to the magnetic field of the magnet. As a voltage is induced across the wire loop, an electrical current in the form of an electron flow starts to flow around the loop generating electricity.

But what if instead of a single individual loop of wire as shown, we had many loops wound together on the same former to form a coil of wire, much more voltage and therefore current could be generated for the same amount of magnetic flux.

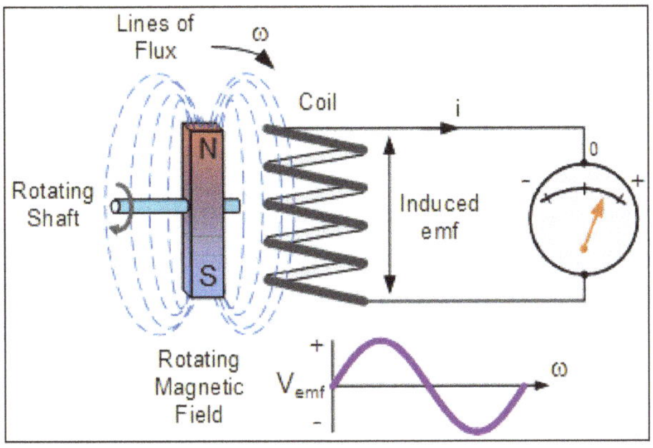

This is because the magnetic flux cuts across more wire producing a greater emf and this is the basic principal of Faraday's law of electromagnetic induction and an AC generator uses this principal to convert a mechanical energy such as the rotation from a wind turbine or hydro turbine, into electrical energy producing a sinusoidal waveform.

So we can see that there are three main requirements for electrical generation and these are:

- A coil or set of conductors.
- A magnetic field system.
- Relative motion between the conductors and field.

Then the faster the coil of wire rotates, the greater the rate of change by which the magnetic flux is cut by the coil and the greater is the induced emf within the coil. Similarly, if the magnetic field is made stronger, the induced emf will increase for the same rotational speed. Thus, emf $\propto \Phi n$. Where, "Φ" is the magnetic-field flux and "n" is the speed of rotation. Also, the polarity of the generated voltage depends on the direction of the magnetic lines of flux and the direction of movement of the conductor.

There are two basic types of electrical generator and alternator for that matter: the permanent-magnet generator and the wound-field generator with both types consisting of two main parts: the Stator and the Rotor.

The stator is the "stationary" (hence its name) part of the machine and can have either a set of electrical windings producing an electromagnet or a set of permanent magnets within its design. The rotor is the part of the machine that "rotates". Again, the rotor can have output coils that rotate or permanent magnets. Generally, generators and alternators used for wind turbine generators are defined by how they make generate their magnetism, either electromagnets or permanent magnets.

There are no real advantages and disadvantages of both types. Most residential wind turbine generators on the market use permanent magnets within their turbine generator design, and which creates the required magnetic field with the rotation of the machine, although some do use electromagnetic coils.

These high strength magnets are usually made from *rare earth* materials such as neodymium iron (NdFe), or samarium cobalt (SmCo) eliminating the need for the field windings to provide a constant magnetic field, leading to a simpler, more rugged construction. Wound field windings have the advantage of matching their magnetism (and therefore power) with the varying wind speed but require an external energy source to generate the required magnetic field.

We now know that the electrical generator provides a means of energy conversion between the mechanical torque generated by the rotor blades, called the prime mover, and some electrical load. The mechanical connection of the wind turbine generator to the rotor blades is made through a main shaft which can be either a simple direct drive, or by using a gearbox to increase or decrease the generator speed relative to the rotational speed of the blades.

The use of a gearbox allows for better matching of the generator speed to that of the turbine but the disadvantage of using a gearbox is that as a mechanical component it is subjected to wear and tear reducing the efficiency of the system. Direct drive however may be more simple and efficient, but the generators rotor shaft and bearings are subjected to the full weight and rotational force of the rotor blades.

Wind Turbine Generator Output Curve

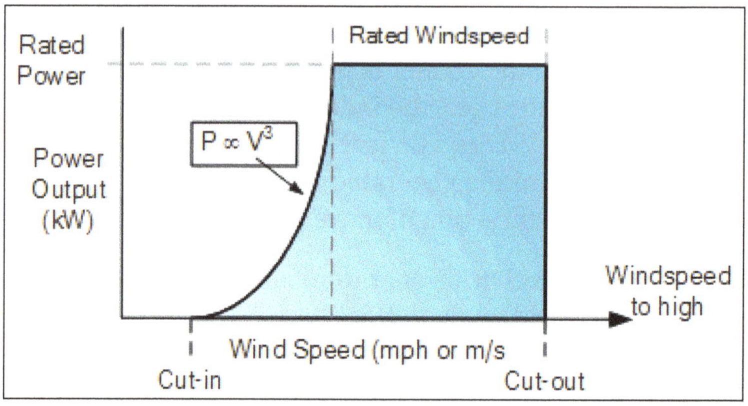

So the type of wind turbine generator required for a particular location depends upon the energy contained in the wind and the characteristics of the electrical machine itself. All wind turbines have certain characteristics related to wind speed.

The generator (or alternator) will not produce output power until its rotational speed is above its cut-in wind speed where the force of the wind on the rotor blades is enough to overcome friction and the rotor blades accelerate enough for the generator to begin producing usable power.

Above this cut-in speed, the generator should generate power proportional to the wind speed cubed ($K.V^3$) until it reaches its maximum rated power output as shown.

Above this rated speed, the wind loads on the rotor blades will be approaching the maximum strength of the electrical machine, and the generator will be producing its maximum or rated power output as the rated wind speed window will have been reached. If the wind speed continues to increase, the wind turbine generator would stop at its cut-out point to prevent mechanical and electrical damage, resulting in zero electrical generation. The application of a brake to stop the generator for damaging itself can be either a mechanical governor or electrical speed sensor.

Buying a wind turbine generator such as the Windmax HY1000 to produce wind energy is not easy and there are a lot of factors to take into account. Price is only one of them. Be sure to choose an electrical machine that meets your needs. If you are installing a grid-connected system, choose an AC mains voltage generator. If you are installing a battery-based system, look for a battery-charging DC generator. Also consider the mechanical design of a generator such as size and weight, operating speed and protection from the environment as it will spend all of its life mounted at the top of a pole or tower.

Energy Collection and Storage

Due to the finite nature of non-renewable energy sources and the increasingly pronounced effects of climate change, securing alternative power sources is becoming a huge issue.

One solution is wind turbines which convert the kinetic energy of the wind into electric energy for consumption. Wind turbines recover the kinetic energy of the moving air by utilizing propeller-like blades, which are turned by wind. The power is transmitted via a shaft to a generator which then converts it into electrical energy. Typically, a group of wind turbines will be installed in the same location known as a 'farm'.

Average sized onshore wind turbines can produce 2.5 to 3 MW of power, offshore wind turbines can produce around 3.6 MW. To put that into perspective, a single offshore turbine can power more than 3,300 average EU households.

Onshore wind has the lowest average levelized cost of all renewable energy sources with an average value of £62/MWh. Power can be generated 24 hours a day, but requires a wind speed of at least 13 mph for utility scale turbines so windy areas of the world are obviously better suited. Offshore locations where winds are stronger and more persistent are ideal locations for wind farms. However, putting farms offshore presents complications in their construction/maintenance and in distributing the power via subsea cables.

How is Wind Power Distributed?

Electricity generated from a wind farm will travel to a transmission substation, where it is stepped up to a high voltage in the region of 150-800 kV. It is then distributed along the electricity grid power lines to the consumer.

Wind is a form of solar energy, the result of uneven heating of the earth's atmosphere by the sun and it is a relatively variable power source. The amount of power generated varies greatly at hourly, daily or seasonal timescales which means that often the supply of electricity will outweigh the demand. In a regular wind farm configuration, the power is distributed straight onto the electrical power grid. With no energy storage capability, this requires the turbines to be slowed to sub-optimal speeds when more energy is produced than is required.

How can Wind Energy be Stored?

Through several different storage processes, excess energy can be stored to be used during periods of lower wind or higher demand.

Battery Storage

Electrical batteries are commonly used in solar energy applications and can be used to store wind generated power. Lead acid batteries are a suitable choice as they are well suited to trickle charging and have a high electrical output charging efficiency.

Compressed Air Storage

Wind turbines can use excess power to compress air, this is usually stored in large above-ground tanks or in underground caverns. When required the compressed air can be used through direct expansion into a compressed air motor. It can also be injected in an internal combustion turbine, where it is burnt with fuel to provide mechanical energy which then powers a generator.

Hydrogen Fuel Cells

Hydrogen fuel cells can also be used to store excess energy. A hydrogen generator is used to electrolyse water using power generated from the wind turbine, storing the resulting hydrogen and converting it back to electricity using a fuel cell power system when needed.

Pumped Storage

Pumped storage is associated with hydroelectric power generation but is yet to be used with wind power generation. Water could theoretically be pumped up to an elevated reservoir utilizing excess generated power and then be used to drive a water turbine when required. The technology is proven and has been used for centuries, giving a relatively high overall efficiency of 70%. Existing hydroelectric power plants could be utilized if they are in an area suitable for a wind farm.

Horizontal-axis Wind Turbines

Horizontal-axis wind turbines (HAWT) have the main rotor shaft and electrical generator at the top of a tower, and may be pointed into or out of the wind. Small turbines are pointed by a simple wind vane, while large turbines generally use a wind sensor coupled with a servo motor. Most have a gearbox, which turns the slow rotation of the blades into a quicker rotation that is more suitable to drive an electrical generator.

Parts of the Wind Turbine

Blades

- The lifting style wind turbine blade. These are the most efficiently designed, especially for capturing energy of strong, fast winds. Some European companies actually manufacture a single blade turbine.
- The drag style wind turbine blade, most popularly used for water mills, as seen in the old Dutch windmills. The blades are flattened plates which catch the wind. These are poorly designed for capturing the energy of heightened winds.

The rotor is designed aerodynamically to capture the maximum surface area of wind in order to spin the most ergonomically. The blades are lightweight, durable and corrosion-resistant material. The best materials are composites of fiberglass and reinforced plastic.

A gear box magnifies or amplifies the energy output of the rotor. The gear box is situated directly between the rotor and the generator. A rotor rotates the generator (which is protected by a nacelle), as directed by the tailvane.

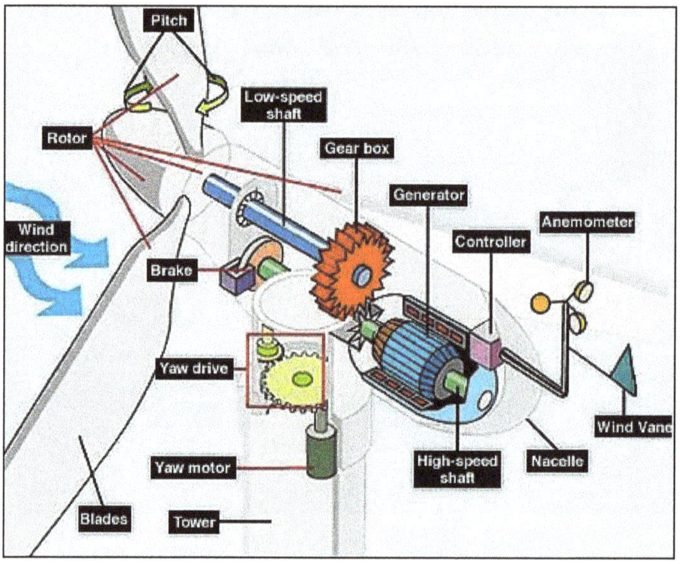

Parts of Wind Turbines.

The generator produces electricity from the rotation of the rotor. Generators come in various sizes, relative to the output you wish to generate. The nacelle is the housing or enclosure that seals and protects the generator and gear box from the elements. It is easily removed for maintenance of the wind. The tailvane directs the turbine to gather maximum wind energy.

Type of Wind the HAWT Works Best in

In general, annual average wind speeds of 5 meters per second (11 mph) are required for gridconnected applications. Annual average wind speeds of 3 to 4 m/s (7-9 mph) may be adequate for non-connected electrical and mechanical applications such as battery charging and water pumping. Wind resources exceeding this speed are available in many parts of the world. A useful way to evaluate the wind resource available at a potential site is the wind power density.

Typical Lifespan

The lifespan of a modern turbine is pegged at around 120,000 hours or 20-25 years, however, they are not maintenance free. As they contain moving components, some parts will need to be replaced during their working life. Throughout research, the cost of maintenance and parts replacement is around the 1 cent USD/AU per kWh or 1.5 to 2 percent annually of the original turbine cost.

HAWT Gearboxes

The gearboxes in the traditional horizontal axis wind turbines currently have an average life span of 1.5 years. Replacing these gearboxes can be extremely costly.

Horizontal Axis Wind Turbine.

Any Power Output Information

Wind turbines for commercial electricity production usual range from 100 kilowatts to 5 megawatts. At the time of writing, the largest wind turbine in the world had a rotor diameter of 126 m (390 feet) and the potential to generate enough electricity for 5000 households. For a 600 kW turbine, the average output is between 1.5 and 2 GWh per year, depending on wind speed. For every kilowatt hour of electricity produced by wind energy or other green means, approximately 1.5 pounds of carbon is prevented from going into the atmosphere if that electricity had been sourced from coal fired power plants. Carbon dioxide is a major contributor to global warming induced climate change.

Vertical-axis Wind Turbine

The Vertical-axis Wind Turbine (VAWT) is the most popular of the turbines that people are adding to make their home a source of renewable energy. While it is not as commonly used as the Horizontal Axis Wind Turbine, they are great for placement at residential locations and more.

Vertical turbines spin on the vertical axis and comes in various shapes sizes and colors. It's movement is similar to a coin spinning on the edge. The main difference between the VAWT and HAWT is the position of blades. In HAWT, blades are on the top, spinning in the air while in VAWT, generator is mounted at the base of the tower and blades are wrapped around the shaft.

Vertical Axis Wind Turbines are designed to be economical and practical, as well as quiet and efficient. They are great for use in residential areas whereas the HAWT is best for use at a business location. There are two different styles of vertical wind turbines out there. One is the Savonius rotor, and the second is the Darrieus model. The first model looks like a 55 gallon drum that is been cut in half with the halves placed onto a rotating shaft. The second model is smaller and looks much like an egg beater. Most of the wind turbines being used today are the Savonius models. We will take a look more in- depth at both of these types of turbines available.

A wind turbine secures air into a hub, which them turns into a generator. The air that passes through the blades of the wind turbine is spun into the generator through rotational momentum. The VAWT, as the turbines are oftener shortened, feature the following qualities:

- Two to three blades with a vertically operating main rotor shaft – the more blades that you have on the unit, the more wind energy it will receive and the more efficiency it will offer.
- Used less frequently than a horizontal wind turbine.
- The position of the blades is different in the VAWT. On this model, the base of the tower holds the generator, and the blades then wrap themselves around the shaft. People use the VAWT because they can be placed closer to the ground, which makes them acceptable and effective for use at a residential location.

- With the vertical axis wind turbine, the rotor shaft is arranged in a vertical pattern.
- The VAWT are easier and more affordable to maintain than horizontal units.
- One complain that some users have with the VAWT is that is creates less wind energy, which may cause a number of different noises to be heart. Turbulent air flow is also a possibility that can shorten the life of the system.
- Installation of the VAWT onto the roof will cause the wind speed to double for maximum wind turbulence and wind energy usage.

Types of Vertical Axis Wind Turbines

As discussed above, there are two different types of VAWTs that you can choose from. While we looked at these types briefly above, now we will take a look at more information about each type and discuss the important factors that you should know. First, let's take a look at the Darrieus wind turbine mode.

Darrieus Wind Turbine

Darrieus Wind Turbine is commonly known as an "Eggbeater" turbine. It was invented by Georges Darrieus in 1931. A Darrieus is a high speed, low torque machine suitable for generating alternating current (AC) electricity. Darrieus generally require manual push therefore some external power source to start turning as the starting torque is very low. Darrieus has two vertically oriented blades revolving around a vertical shaft.

The Darrieus wind turbine offers the following features:

- These eggbeater shaped turbines are great at efficiency, however, they are not as reliable.
- In order to use the Darrieus wind turbine you must have an outside source of power in order to start them.
- It is in your best interest to choose a wind turbine that has at least three blades.
- To support such a wind turbine it is necessary that you have a superstructure which will connect it near the top bearing.

Savonius Wind Turbine

A Savonius vertical-axis wind turbine is a slow rotating, high torque machine with two or more scoops and are used in high-reliability low-efficiency power turbines. Most wind turbines use lift generated by airfoil-shaped blades to drive a rotor, the Savonius uses drag and therefore cannot rotate faster than the approaching wind speed.

Now let's take a look at the second type, which is also the most popular of the two. The Savonius wind turbine is the most popular of the two types. Let's go ahead and look at some of the features these VAWT offer to the homeowner:

- As a drag type of turbine, these units are less efficient.
- When you live in an area that has strong and gusting winds or when you need a unit that self-starts, this is the best type available to you.
- This unit is larger than the Darrieus model.

Savonius vertical axis wind turbine needs to be manually started. The slow speed of Savonius increases cost and produces less efficiency.

Which unit is best for you? There is no right or wrong answer. It is in your best interest to talk to a professional concerning the two types of turbines to determine which is best for you to use.

Advantages of Vertical Axis Wind Turbines

You might be wondering why you would consider using a VAWT instead of a HAWT. There are actually a number of reasons that this decision is made. Let's take a look at some of the advantages that you can enjoy with this type of wind turbine in use at your home:

- You can build your wind turbine close to the ground so if you do not have a suitable rooftop for placement, or if you live where there are hills, ridges, etc. that prohibit the flow of air, they work wonderfully for your needs.
- Since VAWT are mounted closer to the ground they make maintenance easier, reduce the construction costs, are more bird friendly and does not destroy the wildlife.

- You do not need any mechanisms in order to operate the wind turbine.
- Lower wind startup speed.
- The main advantage of VAWT is it does not need to be pointed towards the wind to be effective. In other words, they can be used on the sites with high variable wind direction.
- You can use the wind turbine where tall structures are not allowed.
- VAWT's are quiet, efficient, economical and perfect for residential energy production, especially in urban environments.
- They are cost effective when compare to the HAWTs. It is still best to shop around and check prices before making a purchase, however.
- Many of the turbines are resistant to many of the different weather elements that you may experience. It is imperative to choose a unit that offers this valuable protection and extra durability when you need it the most.

Disadvantages of Vertical Axis Wind Turbines

There are also disadvantages that come with the use of this type of wind turbine. While the many advantages are certainly great, it is imperative that you are aware of the disadvantages. Before deciding which type of wind turbine is best for you it is good idea to take a look at both the pros and the cons. What is right for one person may not be right for you, although it is safe to say that a VAWT is great for almost any residential setting.

Let's take a look at some of the disadvantages of using a VAWT:

- Decreased level of efficiency when compared to the HAWT. The reason for the reduced amount of efficiency is usually due to the drag that occurs within the blades as they rotate.
- You are unable to take advantage of the wind speeds that occur at higher levels.
- VAWT's are very difficult to erect on towers, which means they are installed on base, such as ground or building.

As you can see, there are far more advantages that come your way with the use of a vertical axis wind turbine, but still some disadvantages to be weary of.

VAWT's in general produces less energy as compare to HAWT mainly due to the additional drag that VAWT have – as they rotate into the wind. Both VAWT and HAWT might differ in their appearance and mode of operation, but they generate electricity by converting the wind energy into electrical energy by using a rotating shaft. This is vital information for those looking to install HAWT in their home. Whether they are looking for turbines that will be ideal for when they're sleeping, playing games or entertaining guests, HAWT is the better choice. In other words, wind turns the blades of the wind turbine around the turbines rotor. That rotor is connected to main shaft which in turn is connected to electric generator, which generates the electricity.

If you are looking for a great source of renewable energy that you can use at your residence, it is worth considering a vertical axis wind turbine. These turbines are designed to be used at homes, and while there are both pros and cons, for the most part those who use them appreciate and enjoy the energy that is produced. Before investing into either type of wind turbine, however, it is best to speak to a professional and discuss the pros and the cons of that decision. You'll be glad that you went into the process as an informed consumer. The wind turbine is one of the most effective sources of renewable energy that you can find and you are sure to appreciate all of the wonderful benefits that it is sure to offer to you.

Darrieus Wind Turbine

The Darrieus wind turbines was patented in the United States by G. J. M. Darrieus in 1931. It was reinvented by engineers with the National Research Council of Canada in the early 1970's. Sandia Laboratories built a 5 m diameter Darrieus in 1974, and has been strongly involved with further research on the Darrieus turbine since that time.

Most wind turbines designed for the production of electricity have consisted of a two or three bladed propeller rotating around a horizontal axis. These blades tend to be expensive, high technology items, and the turbine has to be oriented into the wind, another expensive task for the larger machines. These problems have led many researchers in search of simpler and less expensive machines. The variety of such machines seems endless. One that has seen considerable development is the Darrieus wind turbine.

The Darrieus has several attractive features. One is that the machine rotates about a vertical axis, hence does not need to be turned into the wind. Another is that the blades take the shape of a jumping rope experiencing high centrifugal forces. This shape is called troposkein, from the Greek for turning rope.

Since the blade operates in almost pure tension, a relatively light, inexpensive blade is sufficient. Another advantage is that the power train, generator, and controls are all located near ground level, hence are easier to construct and maintain. The efficiency is nearly as good as that of the horizontal axis propeller turbine, so the Darrieus holds considerable promise as a cost effective turbine.

One disadvantage of the Darrieus is that it is not normally self starting. That is, if the turbine has stopped during a period of low wind speeds, it will not usually start when the wind speed increases. Starting is usually accomplished by an induction motor connected to the local utility network.

This is not necessarily a major disadvantage because the same induction motor can be used as an induction generator to supply power to the utility network when the turbine is at operating speed. Induction machines are simple, rugged, and inexpensive, requiring essentially no controls other than a contactor to connect the machine to the utility network. For these reasons, they are seeing wide use as wind turbine generators.

The first large Darrieus constructed was a 230kW machine on Magdalen Island, Quebec, Canada in May, 1977 by Dominion Aluminium Fabricators, Limited of Ontario, Canada. The average power output of this machine was 100 kW over the first year of operation, which is quite good. Then a noise was observed in the gearbox so the machine was stopped forinspection and repairs.

During the inspection process, the brakes were removed, which should have been safe because the turbine was not supposed to be able to self-start. Unfortunately, on July 6, 1978, the turbine started, and without a load or any way of stopping it, went well over the design speed of 38 r/min. The spoilers did not activate properly,and when the turbine reached 68 r/min a guy wire broke, letting the turbine crash to the ground. Perhaps the main lesson learned from this accident was that the Darrieus will sometimes start under unusual gust conditions and that braking systems need to be designed with this fact in mind.

A major design effort on Darrieus turbines has been made by Alcoa. They first designed a 5.5 m diameter machine which would produce about 8 kW of power, but dropped that size in favor of more economical larger machines. Other sizes developed by Alcoa include a 12.8 m diameter (30 to 60 kW), 17 m diameter (60 to 100 kW), and a 25 m diameter (300 or 500 kW depending on the gear ratio).

The Alcoa effort has been plagued by a number of accidents. A 12.8 m diameter machine collapsed at their Pennsylvania facility on March 21, 1980, when its central torque tube started vibrating and eventually buckled when the machine was running above rated speed.

Then in April, 1981, a 25 m machine crashed in the San Gorgonio Pass east of Los Angeles.The machine itself worked properly to a speed well above rated speed, but a software error in the microcomputer controller prevented proper brake application in high winds. When the machine rotational speed reached 60 r/min, well above the rated speed of 41 r/min, a bolt broke and allowed a blade to flare outward and cut one of the guy wires. The machine then crashed to the ground.

Accidents like these are not uncommon in new technology areas, but they are certainly frustrating to the people involved. It appears that the various problems are all solvable, but the string of accidents certainly slowed the deployment of Darrieus turbines as compared with the horizontal axis turbines.

Sandia continued work on the theory of the Darrieus turbine during the 1980s, with the result that the turbine is well understood today. It appears that there is no reason the Darrieus could not be an important contributor to the production of power from the wind. It just needs a large aluminum company that is willing and able to do the aluminum extrusions and possibly wait for several years before seeing a significant return on investment.

Giromill

A Giromill is a Darrieus type vertical axis wind turbine (VAWT) generator. Unlike the more common Savonius VAWT, the Giromill wind turbine uses lift forces generated by vertical aerofoils to convert wind energy into rotational mechanical energy.

The giromill was included in Darrieus's 1927 patent for vertical aerofoil powered vertical axis wind turbines, however development is now starting again on new giromills which take advantage of modern ultra strong light materials to produce turbine blades robust enough to cope with the stresses they are put under.

The giromill is typically powered by two or three vertical aerofoils attached to a central mast by horizontal supports. While it is cheaper and easier to build than a standard Darrieus turbine, it is not as efficient, and requires strong winds (or a motor) to get it to start rotating. They can also sometimes struggle to maintain a steady rate of rotation. However, giromills can work well in turbulent wind conditions and are an affordable option where a standard horizontal axis windmill type turbine would be unsuitable. Once it gets going however, it keeps going well as long as the fan continues to run.

The Cycloturbine

A useful variant of the Giromill is the Cycloturbine which uses a vane to mechanically orient the pitch of the blades for maximum efficiency. This gives the cycloturbine the advantage over other Darrieus wind turbines of self-starting.

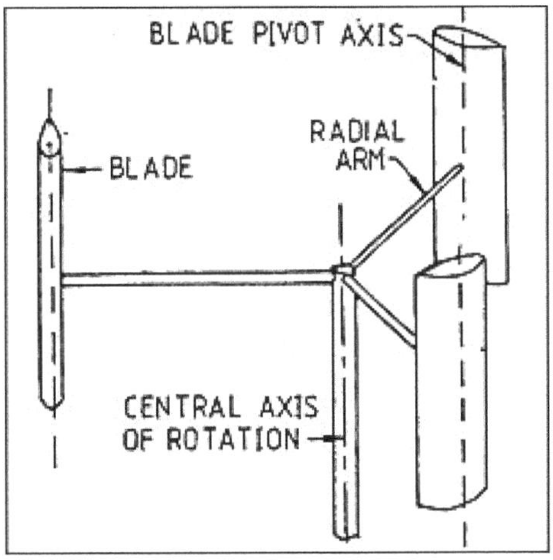

In low winds the blades are pitched flat against the wind generating drag forces and starting the turbine turning. As the rotational speed increases the blades are pitched so that the wind flows across the aerofoils generating lift forces and accelerating the turbine. This mixture of drag and lift (Savonius and Darrieus) makes for an efficient VAWT design, but it is of course more complex, heavy, and expensive to build.

Savonius Wind Turbine

A Savonius is a type of vertical axis wind turbine (VAWT) generator invented in 1922 by Sigurd Johannes Savonius from Finland though similar wind turbine designs had been attempted in previous centuries.

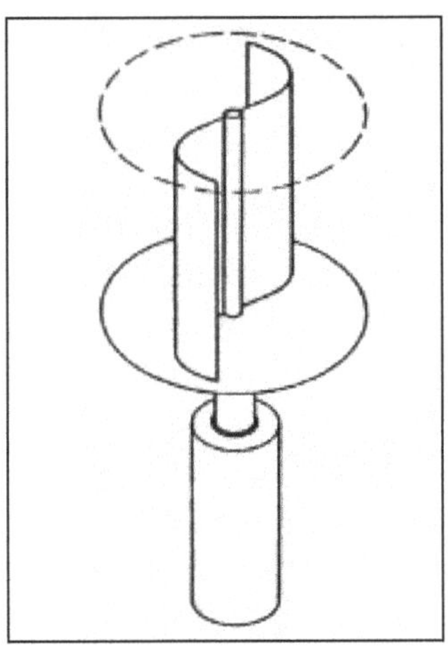

The Savonius is a drag-type VAWT which operates in the same way as a cup anemometer.

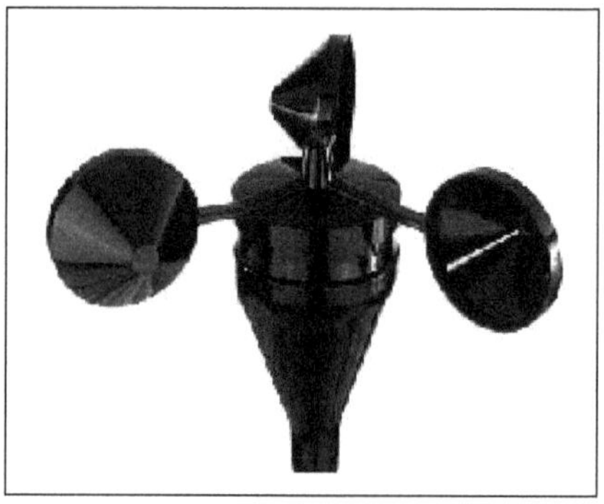

Looking at the anemometer, it has three cups mounted on a rotor which is free to spin. At any time the front of one of the cups will be facing into the wind, while the remaining two cups will be back on to the wind. As the backs of each cup are rounded, they experience much less drag than the front of the cup which is facing into the wind. Therefore the force exerted by the wind on the open cup will be greater than the total force extert-ed on the backs of the other cups and so the rotor will be pushed around (anti-clockwise in the example of the pictured anemometer).

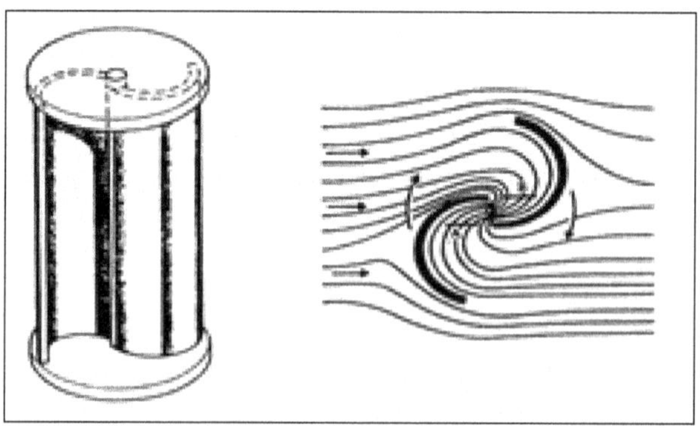

When the rotor has spun around one-third of a revolution, the cup which was open to the wind will now be back on, and one of the previously back on to the wind cups will have spun around on the rotor to offer its open face into the wind. In this way, the rotor will continue to spin for as long as the wind is blowing.

With a Savonius wind turbine it does not matter from which direction the wind is blowing, since there will always be more force exerted on whichever cup has its open face into the wind, and this will push the rotor around. This makes this design of wind turbine ideal for areas with very turbulent wind.

Unfortunately, Savonius wind turbines typically only have an efficiency of around 15% – i.e. just 15% of the wind energy hitting the rotor is turned into rotational mechanical energy. (This is much less than can be achieved with a Darrieus wind turbine which uses lift rather than drag.)

The speed of the cups of a cup anemometer (and a Savonius wind turbine) cannot rotate faster than the speed of the wind they are in and so they have a tip speed ratio (TSR) of 1 or below. This means that a Savonius type vertical axis wind turbines will turn slowly but generate high torque. Therefore Savonius turbines are not ideal for electricity generation since turbine generators need to be turned at hundreds of RPM to generate high voltages and currents. A gearbox can be employed to reduce the torque and increase the RPM of the generator, but that leaves the Savonius requiring a stronger wind to get spinning meaning it may not be able to self-start.

Savonius wind turbines are best suited to applications such as pumping water and grinding grain for which slow rotation and high torque are desireable. They are often used in the third world in exactly these ways with wind turbines made from an old oil barrel (or barrels) welded to a pole which is passed through bearings taken from an old vehicle.

Because of the high torque yield of a Savonius wind turbine, the bearings used must be very sturdy and may require servicing every couple of years.

Savonius Wind Turbines for Electricity Generation

Though not ideal, Savonius wind turbines can and are used in electricity generation – with the benefit that they continue to generate electricity in the strongest winds without being damaged, they are very quiet, and they are relatively easy to make.

Standard HAWT (windmill) type wind turbines are unsuitable in locations with strong turbulent winds, so Savonius wind turbines can sometimes be the best option. They can also be mounted relatively safely at ground level as they spin much much slower than the speed of the tips of an HAWT.

Small Wind Turbine

Small wind turbines (SWTs) are a distinct and separate group of devices developed within the wind energy sector. According to the IEC 61400-2 standard, SWTs are characterized by a rotor area of <200 m^2 and rated power below 50 kW. Wind power plants in this category are generally designed for small and individual customers such as households, farms, weather stations, road signalization, and advertising systems. SWTs offer a promising alternative for many remote electrical uses where, given a set of site evaluation criteria, the wind resources can be identified as beneficial, both as stand-alone applications and in combination with other energy conversion technologies such as photovoltaic, small hydro or diesel engines.

The quantity of SWTs operating worldwide grows every year. In 2012, the total number of such devices was approximately 800,000 worldwide with the growth of about 10%. The majority of SWTs (about 70%) are located in China, where the highest number of new installed units in 2012 was also noted. The second biggest market of SWTs is USA, where around 155,000 SWTs are operating at the time this document is prepared. In Europe, the leader is the United Kingdom: 23,500 units, followed by Germany: 10,000 units, Spain: 7020 units, and Poland: 3200 units. Total SWT generation capacity installed in 2012 was equal to around 678 MW (576 MW in 2011). The majority of world's capacity (85%) belongs to three countries: China (274 MW), USA (216 MW), and UK

(83.7 MW). Unfortunately, developing countries play a minor role in small wind turbine industry. Electrical capacity growth in 2013 was small, with just 90 MW installed across Africa, for a cumulative total of 1255 MW. It is exceptionally regrettable considering enormous wind power potential (best around the coasts and in the eastern highlands of the African continent). A global forecast concerning SWTs installed capacity in years 2009–2020 is presented in figure below:

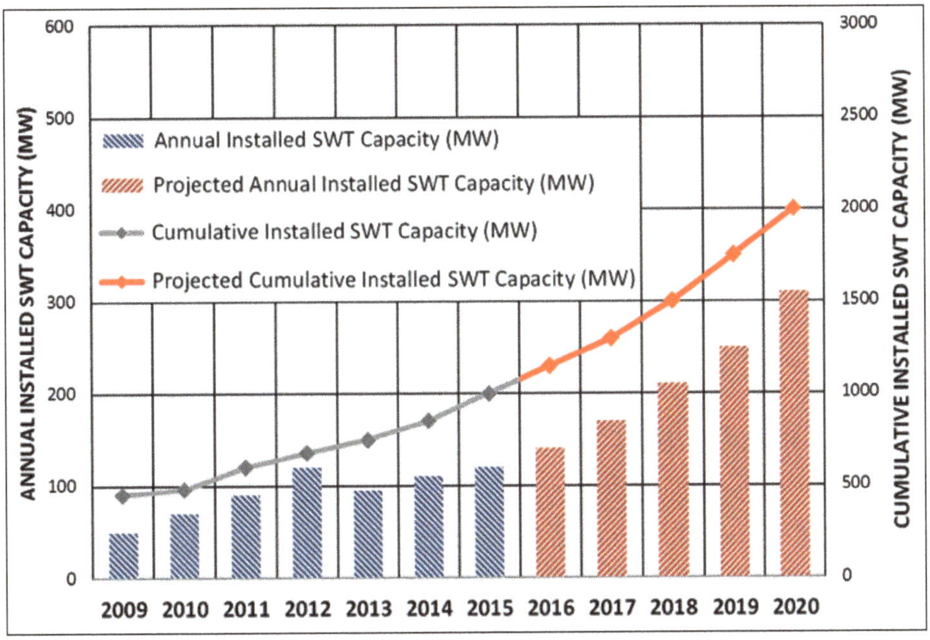

SWT installed capacity world market forecast for 2009–2020.

The development and dissemination of SWTs involves great expectations in the field of eco-energy production. Some opinions suggest that without the dissemination of SWTs, the fulfillment of legal requirements for energy efficiency and energy production from renewable sources will be relatively difficult. In particular, in the developing countries, small wind turbine sector can efficiently contribute to provide electricity to millions of people in rural areas. In order to create a positive outcome, a big challenge awaits not only the authors of laws supporting investments in SWTs, but also engineers and scientists who should propose design solutions addressing the real issues found in the small wind turbine energy sector (aerodynamics: ill-optimized blade often using trends and observations from large HAWTs, poorly addressed issue of local wind resources abundance; structural: different nature of loads experienced by SWTs, little care given to weight optimization; conversion/control: seldom usage of active regulation methods, such as rpm control to expand operational margin of the SWTs, little attention given to match turbine's mechanical power capability with that of a generator; economic: lack of cost optimized SWTs).

The wide prevalence of SWTs and the emergence of the so-called "prosumers" within the electrical grid (the Smart Grid concept) are believed to be one of the biggest factors changing the way that the power companies will deliver service over the next decade. This qualitative change may result in a reduction of transmission losses and make the electricity infrastructure more flexible and secure. The specialists agree that work on the design and architecture of the future grid is as important as the work on the technologies and products that would realize a smart grid vision, for example, SWTs being a part of electricity network. The above motivates to devote more effort to the field of modern SWTs development.

In the small wind turbine market, the return on investment (ROI) is one of the most important factors determining the turbine's validity. Having the above in mind, it has become a challenge for many designers and research facilities to develop a small wind turbine design which would be competitive with other renewable energy sources. For this to be possible, it would have to incorporate a number of factors: high efficiency, sufficient longevity, low installation, and maintenance costs. Having above in mind, it has been concluded that SWTs should be characterized by the lowest possible price, while maintaining relatively high efficiency as well as satisfying reliability and maintenance parameters.

SWT's Economic Evaluation

The investment in SWTs is more often made by private individuals in order to partially fulfill residential household demand for electricity or to produce hot tap water. However, practice shows that a large portion of those who have decided to build a backyard wind turbine is disappointed with the actual amount of energy produced by it. This is mainly due to the promotion of inaccurate, overstated, or even incorrect information on the projected turbine's power output by manufacturers or installers. This reduces public confidence in the legitimacy of this sort of economic investment, thus slowing down the development of this industry sector.

Numerical weather prediction (NWP) simulations over the span of year 2013, covering the whole area of Poland, were carried out. The purpose of these forecasts was to establish a reliable and accurate wind resource atlas for approximate AEP of small wind turbine systems. Long-term wind speed forecasting has been used with success for utility-scaled wind farms. This accounts not only for planned sites but also operational power plants, where weather forecasting is used constantly for setting up advanced control schemes or predicting suitable time windows for planned maintenance.

Calculation of meteorological parameters was performed using the non-hydrostatic mesoscale Weather Research & Forecasting (WRF) model. WRF is designed both for operational forecasting and atmospheric research use. It enables an atmospheric phenomena simulation for scales ranging from thousands of kilometers to single meters. The model contains several capabilities, important from the research point of view, such as: 3D data control with initial data and model results correction possibilities using aerological, radar and satellite measurements; multilayer ground modeling;

humidity cycle; vegetation; cloud cover and precipitation parameterization including water phase state; radiation transport processes; and boundary layer with turbulent vertical transport. These modules are responsible for the parameterization of physical phenomena occurring in the atmosphere. The boundary layer area of intensive surface radiative forcing and mechanical forcing is an important part of the simulated phenomena. Mechanical forcing is determined by orography, roughness, and cover. Radiative forcing is defined by albedo and thermal capacity of the surface. The parameters are determined by the ratio of land to water area; vegetation and its status; irradiation depending on location and inclination. Determination of these factors was of importance because the boundary layer close to the surface is the area associated with the presented studies.

The data used in the study covered the period of 1 year. The input data of the WRF came from the archive of the global forecast system (GFS). These were obtained from the National SOO Science and Training Resource Centre (NWS SOO/STRC). The model was run for each day and for each main synoptic term. Two grids with spatial resolution of 36 and 12 km were used. Forecast modeling time was set equal to 24 h with a 1 h data sampling rate. Results for each 24-h period were obtained in a five-dimensional set of prognostic parameters fields. This set includes, among others: pressure values, geopotential, temperature, and three-dimensional wind fields. Wind speed data averaged over a 60 min period have been said to be useful for long-term energy yield predictions for SWTs.

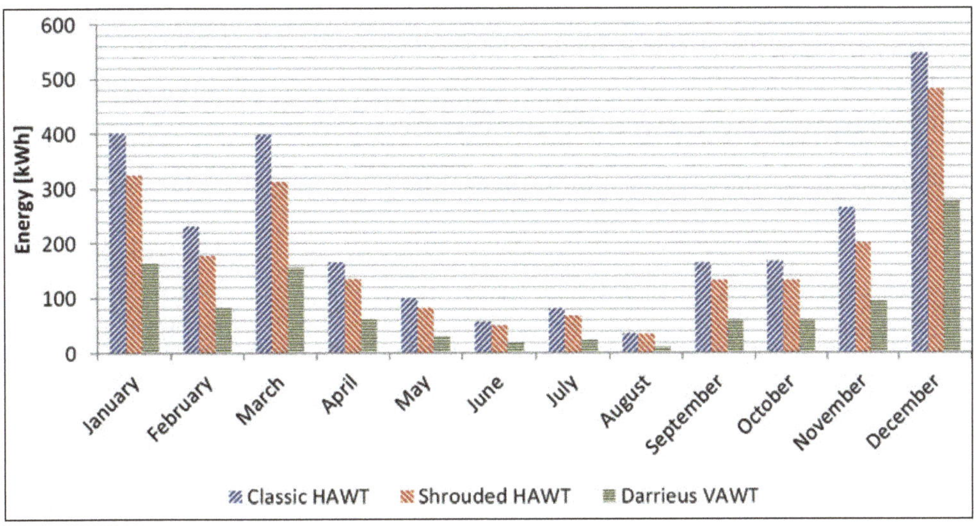

Monthly energy output for the chosen turbines.

A randomly chosen location in model domain was a rural region neighboring a small town called Sokolow Malopolski in the Voivodship of Sub-Carpathia. It is characterized with rather typical rural wind speed conditions. Many of the houses built in the region are solitary and surrounded by open grass or farm fields, thus producing a suspected low ground roughness rate. Such households could potentially benefit from an auxiliary power or heating source in the form of a small wind turbine.

Calculations of the power output were performed using the following formula:

$$E\sum_{i=1}^{n}\begin{cases}\frac{1}{2}A\rho i\eta(v_i)v_i^3 h & P_r > \frac{1}{2}A\rho i\eta(v_i)v_i^3 \\ P_r h & P_r \geq \frac{1}{2}A\rho i\eta(v_i)v_i^3\end{cases}$$

where: E is the estimated total energy generated throughout the test period, n is the number of samples, h is the sample duration (1 h), v_i is the wind speed for sample i, A is the swept area of a given turbine, ρ_i is the air density for sample i, $\eta(v_i)$ wind turbine efficiency at given tip-speed ratio (TSR) and P_r is the turbine rated power.

HAWT, VAWT and DAWT Comparison

Three types of SWTs were chosen to be compared in terms of their productivity and cost-effectiveness. The first is a typical three-bladed HAWT design. The second is an aerodynamically complex, shrouded HAWT. The third is the Darrieus-type VAWT. All the turbine characteristics, such as power curves, price, swept area, were taken from very popular, commercially proven designs. For the results to be meaningful, all the chosen turbines have a similar rated power of around 3 kW.

The outcome of the performed calculations is shown in figure and compared in table. When estimating the investment return period, the price for energy supplied (power supply) was established at 0.2 $/kWh. Maintenance and any additional costs associated with turbine operation are assumed to be the same for all three systems and are therefore neglected.

Table: Chosen designs economic comparison.

	Energy produced (kWh/year)	ROI (years)
Classic HAWT	2610	28+
Shrouded HAWT	2126	31+
Darrieus VAWT	1036	137+

The difference in monthly power generation is clearly visible with the majority of energy being produced during the windy winter months. Any needed maintenance or routine inspections should be carried out during calm summer months. Such a discrepancy in electricity generation raises the problem of continuous power supply.

For the chosen localization, the classic three-bladed HAWT proved to be the most cost-effective design of those taken into account, even though it had the lowest rated power. In theory, the advanced construction of the shrouded HAWT turbine allowed for a very high theoretical efficiency. In practice, because of its small rotor diameter it proved to be slightly inferior to the classic layout with a larger rotor. Moreover, the shroud is likely to put additional strain on the tower due to added mass and increased aerodynamic drag, possibly increasing the maintenance costs. Snowfall and icing of the shroud could lower the operational availability and introduce the need for deicing. The chosen Darrieus vertical wind turbine was the worst design in the terms of cost-effectiveness. Very high system costs and low efficiency in relation to rotor size led to the return period being unacceptably long.

Investigation of Parameters Influencing the Efficiency of SWTs

Three cases will be carried out in order to find the factors which have the highest impact on the estimated electrical power generation. Case one investigates the difference between estimated one with a theoretical, constant efficiency turbine and a popular, commercially available, wind turbine using its power curve. The second case estimates the generated amount of energy based on three types of wind data for a randomly chosen site. In this case, calculations were performed using a yearly mean wind speed, discretized meteorological data and the Weibull wind speed distribution function. The third case shows the difference of estimated energy production for three tower heights: 10, 20, and 30 m. Calculations of the power outputs were performed using formula.

The turbine chosen for the analyses is the Evance R9000, which is a popular small wind turbine design with a high market share, especially in the UK. It is a 5 kW turbine, with a 5.5-m-radius rotor and a typical, of HAWTs, power curve. Factors such as maintenance and down time were not taken into account. That is due to the fact that the goal for these calculations was to estimate small wind turbine sensitivity to operating parameters such as a tower height, efficiency, and wind data type with which the AEP forecast is modeled.

The estimated energy produced by the Evance R9000 amounted to 7551 kWh, assuming energy cost of 0.2 €/kWh, that is an equivalent of 1510 €. For a 5-year ROI period, without any maintenance or down time, the turbine would have to be priced at around 7500 €; however, in reality, such solutions are roughly priced at 36,500 €. Another aspect worth mentioning is that a yearly demand for energy of typical household is estimated at around 3–4 MWh. Thus, in theory, it would be possible to achieve energy independence. In practice, it would be nearly impossible.

Another very interesting way to present the obtained results is to draw an energy density function on top of the occurring wind speeds. Figure shows, in blue, the wind speed occurrence rates and, in red, the amount of energy generated from a specific wind speed by the Evance R9000 turbine during the whole year. The shift of the energy density function to the right in respect to wind speed curve is a direct result of the power equation, proportional wind speed cubed equation,

$$E\sum_{i=1}^{n}\begin{cases}\frac{1}{2}A\rho i\eta(v_i)v_i^3 h & \mathrm{P_r} > \frac{1}{2}A\rho i\eta(v_i)v_i^3\\ \mathrm{P_r}\, h & \mathrm{P_r} \geq \frac{1}{2}A\rho i\eta(v_i)v_i^3\end{cases}.$$

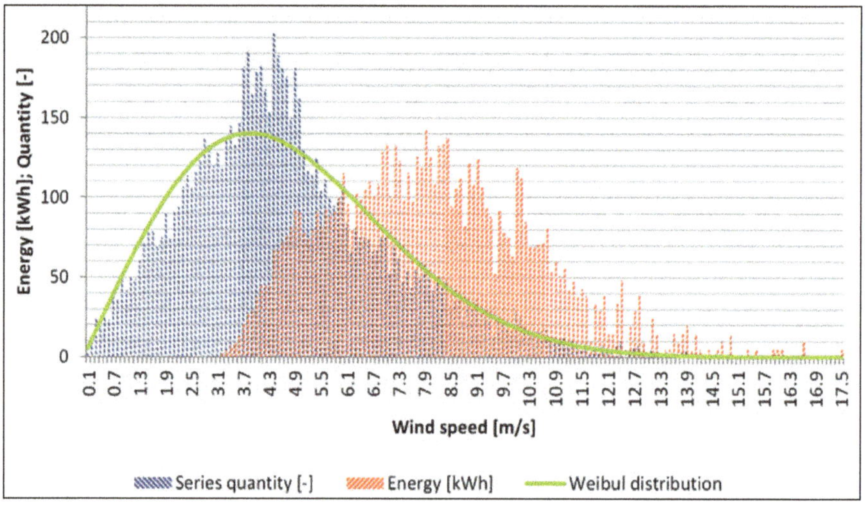

Wind speed distribution, energy distribution, and Weibull function fit (c = 5379; k = 2023).

Using Matlab computer software, the maximum likelihood estimate of the Weibull parameters has been established with the wlbfit function. The resultant Weibull probability density function is shown in figure. Additionally, an arithmetic mean of the wind speed for the given location has been calculated and is equal to around 4.77 m/s. The yearly power generated has been again calculated using the obtained Weibull function and the mean wind speed with the standard Evance9000 power curve. A comparison of the estimated power production based on three different wind data types is presented in figure. It is clearly visible that estimating the amount of wind power generation using an arithmetic mean of the wind speed is highly inaccurate. The predicted amount was 4420 kWh which is almost 70% off the amount of energy predicted using the direct method with accurate wind speed data. Energy production estimated with the Weibull function equaled 7932 kWh, which is 5% different from the result obtained using the direct method. This difference is small enough to prove the validity of using the Weibull or Rayleigh distributions for estimating small wind turbine energy production.

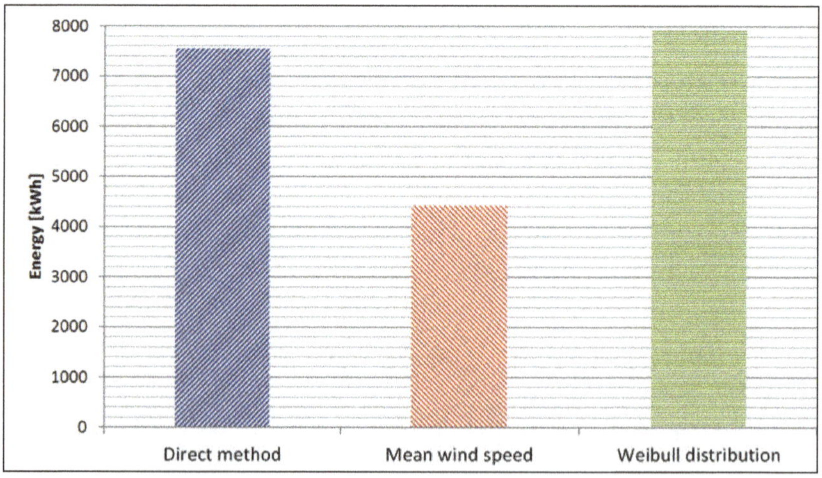

Energy production estimates based on three different wind speed data types.

Five different calculations of estimated energy output were performed with five different small wind turbine characteristics. The first was performed with an artificial, constant value of 33%. Three characteristics were based on the Evance R9000 power curve including the original curve, one which is offset by +5% and another which is offset by −5% from the original distribution. The fifth one is a hypothetically proposed, small wind turbine curve which has increased efficiency in low wind speeds—STOW. The results of the performed calculations are presented in figure.

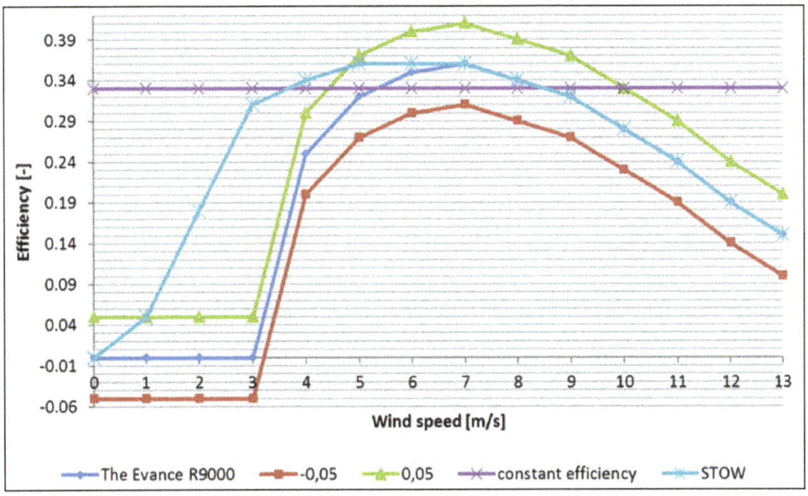

Efficiency curves used for the energy calculations.

The final investigated small wind turbine parameter was the tower height. Wind speed data were collected at 10 m height. Using the power law with the terrain dependent parameter α = 1/7, the wind speeds for the whole year have been recalculated for 20 and 30 m heights. The obtained wind speeds were used to estimate the produced power with the standard power curve and are presented in figure.

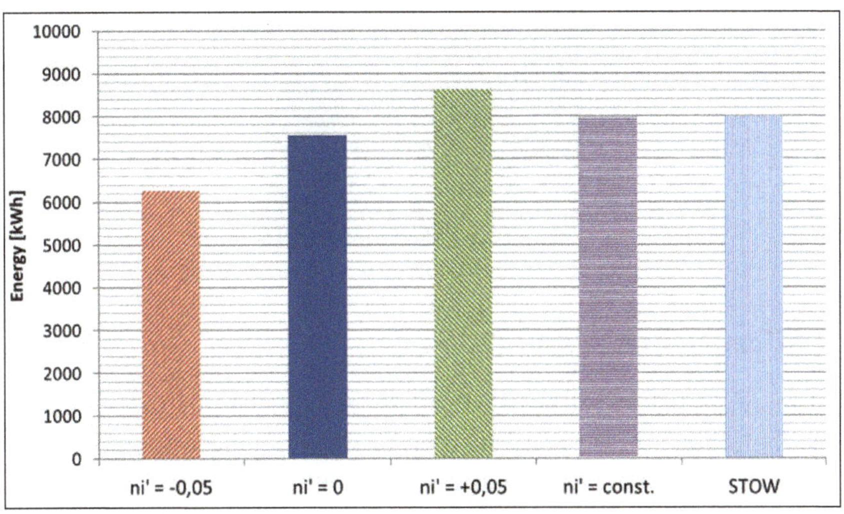

Energy produced by a 5 kW rated turbine with different efficiency curves; the series are tied to the curves.

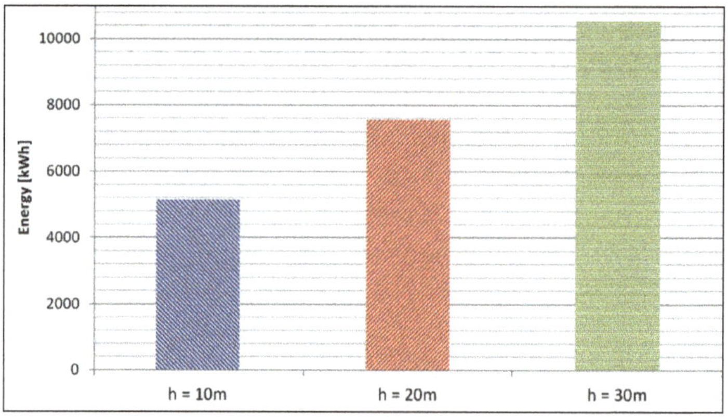

Estimated energy generated for different tower height variants (10, 20, and 30 m)—wind speed obtained with the empirical power law.

Micro-wind Turbine

Micro-wind turbines are used in micro-wind generation and are much smaller in scale than those used in conventional wind generation making them more suitable for residential energy production. Micro-wind generation is a method of microgeneration that uses the flow of wind energy to produce electricity for a house or farm. Broadly speaking, there are two types of wind turbines that can be installed: vertical axis wind turbines and horizontal axis wind turbines.

An example of micro wind turbine used for a home.

What is a Micro-wind System?

The installation of a micro-wind turbine usually consists of the turbine and an inverter. Wind causes the blades of the wind turbine to rotate, generating mechanical energy. The mechanical energy from the rotation is converted to direct current (DC) in the turbine and using the inverter, is converted to alternating current (AC). The inverter output is connected to a breaker panel where the electricity can be shared among the electrical equipment in the home. Excess electricity can be exported from the home to the electrical grid using a bidirectional meter and credits will be provided accordingly by the retailer based on the electric current tariff for electricity.

Installation Requirements

The approvals and permits required for micro-wind generation are the same as what would be required for other forms of microgeneration. These approvals and permits are normally obtained from the organization of the judicial government responsible for electric utilities. This will normally be obtained prior to the installation at which point the retailer will provide a home evaluation to ensure the qualifications required for installation are met. Additional approvals may also need to be attained such as in Alberta, Canada where the installations of these turbines must comply with air navigation or aeronautical safety.

Factors Affecting Generation Performance

The generation of electricity is mostly based on the rotational wind speeds of the wind turbines. Certain geographic locations are more suitable for producing electricity compared to others. Depending on the amount of wind that can be obtained from a region, the generation can vary. Another factor that can affect the performance of wind turbines is the obstructions in the area of the installed turbine. Obstructions from trees or other buildings will hinder the turbines from producing at its optimum capacity.

Typhoon-proof Wind Turbine

While wind energy is rapidly increasing its market share across the world, wind turbines are not able to be constructed everywhere that they might be needed. A perfect example of this is Japan, where a traditional wind turbine would get damaged by typhoons. After the Fukushima disaster, though, one Japanese engineer committed himself to building a turbine specifically for Japan that can operate just fine within hurricane-force winds.

The "typhoon turbine" as it is known works via the Magnus effect, where a spinning object directs air around it faster on one side than on the other. This turbine uses three Magnus effect-driven cylinders with a blade on each one, which allows the turbine to harvest energy no matter how high the wind speeds are. The problem with hurricanes and typhoons isn't just the wind, but also what the wind blows around. While there is no mention of its impact resistance it certainly looks like it has been built as robustly as possible.

Hopefully this turbine is able to catch on in Japan so they can reduce their reliance on other types of energy. Wind energy has been getting incredibly popular lately, including among hikers who carry a portable wind generator, and even among people with just a few pieces of scrap material.

References

- Windturbines: conserve-energy-future.com, Retrieved 21 May, 2019
- Wind-turbine-generator, wind-energy: alternative-energy-tutorials.com, Retrieved 8 January, 2019
- Horizontal-axis-wind-turbines-hawt: turbinesinfo.com, Retrieved 13 May, 2019
- Verticalaxiswindturbines: conserve-energy-future.com, Retrieved 25 February, 2019
- Darrieus-wind-turbines: turbinesinfo.com, Retrieved 16 January, 2019
- Giromill-darrieus-wind-turbines, wind, wordpress: reuk.co.uk, Retrieved 29 March, 2019
- Savonius-wind-turbines, wind, wordpress: reuk.co.uk, Retrieved 30 April, 2019
- Small-wind-turbines-specification-design-and-economic-evaluation, wind-turbines-design-control-and-applications: intechopen.com, Retrieved 29 June, 2019

Chapter 3

Wind Turbine: Design, Aerodynamics and Measurements

Wind turbine design includes rotor, direct current generator or alternating current generator which are used to define the specifications of a wind turbine for extracting wind energy. Aerodynamic modeling, classic barrel design of Savonius, theory of wind turbine and Betz coefficient, etc. are some of the concepts that fall under its domain. This chapter closely examines these concepts of wind turbine to provide an extensive understanding of the subject.

Wind Turbine Design

At the heart of any renewable wind power generation system is the Wind Turbine. Wind turbine designs generally comprise of a rotor, a direct current (DC) generator or an alternating current (AC) alternator which is mounted on a tower high above the ground.

So how are wind turbines designed to produce electricity?. In its simplest terms, a wind turbine is the opposite to a house or desktop fan. The fan uses electricity from the mains grid to rotate and circulate the air, making wind. Wind turbine designs on the other hand use the force of the wind to generate electricity. The winds movement spins or rotates the turbines blades, which captures the kinetic energy of the wind and convert this energy into a rotary motion via a shaft to drive a generator and make electricity as shown.

Typical Wind Turbine Generator Design

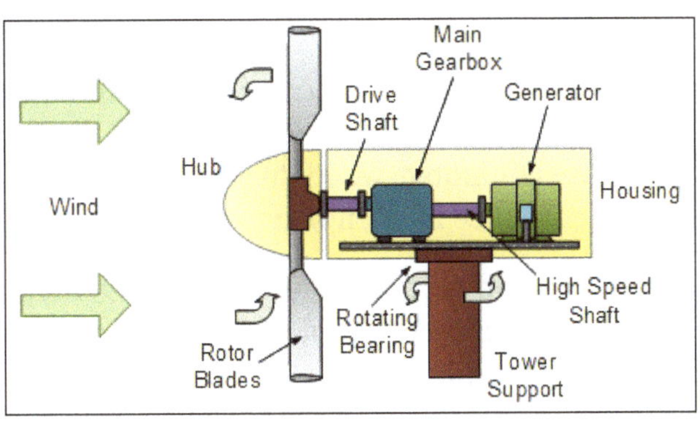

The image above shows the basic components that go to make up a typical wind turbine design. A wind turbine extracts the kinetic energy from the wind by slowing the wind down, and transferring this energy into the spinning shaft so it is important to have a good design. The available power in the wind that is available for harvesting depends on both the wind speed and the area that is swept by the rotating turbine blades. So the faster the wind speed or the larger the rotor blades the more energy can be extracted from the wind. So we can say that wind turbine power production depends on the interaction between the rotor blades and the wind and it is this interaction that is important for a wind turbine design.

To help improve this interaction and therefore increase efficiency two types of wind turbine design are available. The common horizontal axis and the vertical axis design. The horizontal axis wind turbine design catches more wind so the power output is higher than that of a vertical axis wind turbine design. The disadvantage of the horizontal axis design is that the tower required to support the wind turbine is much higher and the design of the rotor blades has to be much better.

A Typical Wind Turbine Blade Design.

The Vertical Axis Turbine or VAWT, is easier to design and maintain but offers lower performance than the horizontal axis types due to the high drag of its simple rotor blade design. Most wind turbines generating electricity today either commercially or domestically are horizontal axis machines so it is these types of wind turbine design we will look at.

The Rotor: This is the main part of a modern wind turbine design that collects the winds energy and transforms it into mechanical power in the form of rotation. The rotor consists of two or more laminated-wood, fibreglass or metal "rotor blades" and a protective hub which rotates (hence its name) around a central axis.

Just like an aeroplane wing, wind turbine blades work by generating lift due to their curved shape. The rotor blades extract part of the kinetic energy from the moving air masses according to the lift principle at a rate determined by the wind speed and the shape of the blades. The net result is a lift force perpendicular to the direction of flow of the air. Then the trick is to design the rotor blade to create the right amount of rotor blade lift and thrust producing optimum deceleration of the air and no more.

Unfortunately the turbines rotor blades do not capture 100% all of the power of the wind as to do so would mean that the air behind the turbines blades would be completely still and therefore not allow any more wind to pass through the blades. The theoretical maximum efficiency that the turbines rotor blades can extract from the wind energy amounts to between 30 and 45% and which is dependant on the following rotor blade variables: Blade Design, Blade Number, Blade Length, Blade Pitch/Angle, Blade Shape, and Blade Materials and Weight to name a few.

Blade Design: Rotor blade designs operate on either the principle of the lift or drag method for extracting energy from the flowing air masses. The lift blade design employs the same principle that enables aeroplanes, kites and birds to fly producing a lifting force which is perpendicular to the direction of motion. The rotor blade is essentially an aerofoil, or wing similar in shape to an aeroplane wing. As the blade cuts through the air, a wind speed and pressure differential is created between the upper and lower surfaces of the blade.

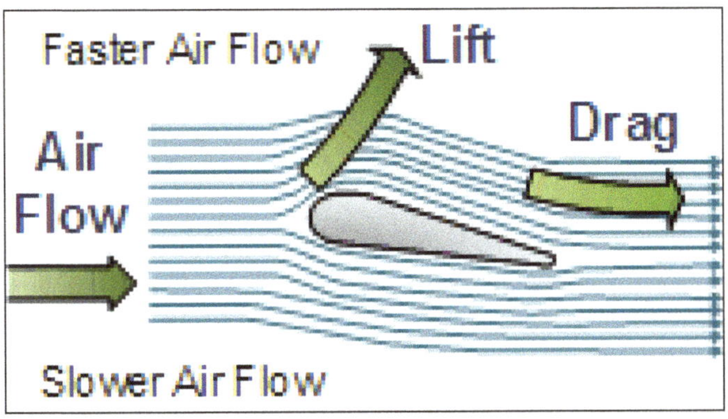

The pressure at the lower surface is greater and thus acts to "lift" the blade upwards, so we want to make this force as big as possible. When the blades are attached to a central rotational axis, like a wind turbine rotor, this lift is translated into a rotational motion.

Opposing this lifting force is a drag force which is parallel to the direction of motion and causes turbulence around the trailing edge of the blade as it cuts through the air. This turbulence has a braking effect on the blade so we want to make this drag force as small as possible. The combination of lift and drag causes the rotor to spin like a propeller.

Drag designs are used more for vertical wind turbine designs which have large cup or curved shaped blades. The wind literally pushes the blades out of the way which are attached to a central shaft. The advantages of drag designed rotor blades is slower rotational speeds and high torque capabilities making them useful for water pumping and farm machinery power. Lift powered wind turbines having a much higher rotational speed than drag types and therefore are well suited for electricity generation.

Blade Number: The number of rotor blades a wind turbine design has is generally determined by the aerodynamic efficiency and cost. The ideal wind turbine would have many thin rotor blades but most horizontal axis wind turbine generators have only one, two or three rotor blades. Increasing the number of rotor blades above three gives only a small increase in rotor efficiency but increases its cost, so more than three blades are usually not required but small high spinning multi-bladed turbine generators are available for home use. Generally, the fewer the number of blades, the less material is needed during manufacturing reducing their overall cost and complexity.

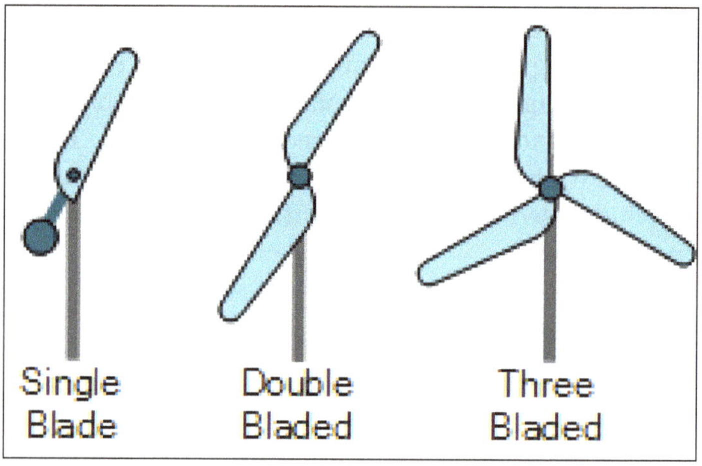

Single bladed rotors have a counter balance weight on the opposite side of the rotor but suffer from high material stress and vibration due to their unsmooth rotational motion of the single blade which must move more rapidly to capture same amount of wind energy. Also with single or even double bladed rotors, most of the available air movement and therefore wind power passes through the unswept cross-sectional area of the turbine without interacting with the rotor decreasing their efficiency.

Multi-bladed rotors on the other hand have a smoother rotational operation and lower noise levels. Slower rotational speeds and torque are possible with multi-bladed designs which reduces the stresses in the drive train, resulting in lower gearbox and generator costs. However, wind turbine designs with many blades or very wide blades will be subject to very large forces in very strong winds which is why most wind turbine designs use three rotor blades.

An Odd or Even Number of Rotor Blades?: A wind turbine design which has an "EVEN" number of rotor blades, 2, 4 or 6, etc, can suffer from stability problems when rotating. This is because each rotor blade has an exact and opposite blade which is located 180° in the opposite direction. As the rotor rotates, the very moment the uppermost blade is pointing vertically upwards (12 o'clock position) the lower most blade is pointing straight down in front of the turbine support tower. The result is that the uppermost blade bends backwards, because it receives the maximum force from the wind, called "thrust loading", while the lower blade passes into the wind free area directly in front of the supporting tower.

This uneven flexing of the turbines rotor blades (uppermost bent in the wind and the lowermost straight) at each vertical alignment produces unwanted forces on the rotor blades and rotor shaft as the two blades flex back and forth as they rotate. For a small rigid aluminium or steel bladed turbine this may not be a problem unlike longer fibreglass reinforced plastic blades.

A wind turbine design which has an "ODD" number of rotor blades (at least three blades) rotates smoother because the gyroscopic and flexing forces are more evenly balanced across the blades increasing the stability of the turbine. The most common odd bladed wind turbine design is that of the three bladed turbine. The power efficiency of a three bladed rotor is slightly above that of a similar sized two bladed rotor and due to the additional blade they can rotate slower reducing wear and tear and noise.

Also, to avoid turbulence and interaction between the adjoining blades, the spacing between each blade of a multi-bladed design and its rotational speed should be big enough so that one blade will not encounter the disturbed, weaker air flow caused by the previous blade passing the same point just before it. Because of this limitation most odd type wind turbines have a maximum of three blades on their rotors and generally rotate at slower speeds.

Generally, three bladed turbine rotors integrate better into the landscape, are more aesthetically appealing and are more aerodynamically efficient than two bladed designs which contributes to the fact that three bladed wind turbines are more dominate in wind power generation market. Although certain manufacturers produce two and six-blade turbines (for sail boats). Other advantages of odd (three) bladed rotors include smoother operation, less noise and fewer bird strikes which compensate for the disadvantage of the higher material costs. Noise level is not affected significantly by the blade count.

Rotor Blade Length: Three factors determine how much kinetic energy can be extracted from the wind by a wind turbine: "the density of the air", "the speed of the wind" and "the area of the rotor". The density of the air depends upon how far above sea level you are while the wind speed is controlled by the weather. However, we can control the rotational area swept by the rotor blades by increasing their length as the size of the rotor determines the amount of kinetic energy a wind turbine is able to capture from the wind.

The rotor blades rotate around a central bearing forming a perfect circle of 360° as it rotates and the area of a circle is given as: $\pi.r^2$. So as the swept area of the rotor increases, the area it covers also increases with the square of the radius. Thus, doubling the length of a turbines blades results in an increase of four times its area which allows it to receive four times as much wind energy. However, this greatly increases the size, weight and ultimately the cost of the wind turbine design.

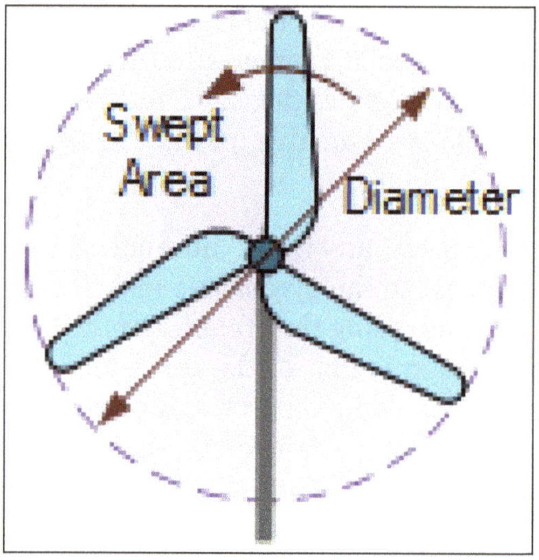

One important aspect of the blade length is the rotational tip-speed of the rotor resulting from the angular velocity. The longer the turbine blade length the faster the rotation of the tip for a given wind speed. Likewise, for a given rotor blade length the higher the wind speed the faster the rotation. So why can we not have a wind turbine design with very longer rotor blades operating in a windy environment producing lots of free electricity from the wind. The answer is that there becomes a point where the length of the rotor blades and the speed velocity of the wind actually reduces the output efficiency of the turbine. This is why many larger wind turbine designs rotate at much slower speeds.

Efficiency is a function of how fast the rotor tip rotates for a given wind speed producing a constant wind speed to tip ratio called the "tip-speed ratio" (λ) which is a dimensionless unit used to maximise the rotor efficiency. In other words, "tip-speed ratio" (TSR) is the ratio of the speed of the rotating blade tip in rpm to the speed of the wind in mph, and a good wind turbine design will determine the rotor power for any combination of wind and rotor speed. The larger this ratio, the faster the rotation of the wind turbine rotor at a given wind speed. The shaft speed that the rotor is fixed too is given in revolutions per minute (rpm) and depends on the tip-speed and the diameter of the turbines blades. A turbines rotational speed is defined as, rpm = wind speed × tip-speed-ratio × 60 / (diameter × π).

If a turbines rotor rotates too slowly, it allows too much wind to pass through undisturbed, and thus does not extract as much as energy as it could. On the other hand, if the rotor blade rotates too quickly, it appears to the wind as one large flat rotating circular disc, which creates large amounts of drag and tip losses slowing the rotor down. Therefore it is important to match the rotational speed of the turbine rotor to a particular wind speed so that the optimum efficiency is obtained. Turbine rotors with fewer blades reach their maximum efficiency at higher tip-speed ratios and generally, three bladed wind turbine designs for electrical generation have a tip speed ratio of between

6 and 8, but will run more smoothly because they have three blades. On the other hand, turbines used for water pumping applications have a lower tip speed ratio of between 1.5 and 2 as they are specially designed for high torque generation at low speeds.

Rotor Blade Pitch/Angle has fixed design wind turbine rotor blades are generally not straight or flat like aeroplane aerofoil wings, but instead have a small twist and taper along their length from the tip to the root to allow for the different rotational speeds along the blade. This twist allows for the blade to absorb the winds energy when the wind is coming at it from different tangential angles and not just straight-on. A straight or flat rotor blade will stop giving lift and may even stop (stall), if the rotor blade is hit by the wind at different angles, called the "angle of attack" especially if this angle of attack is too steep.

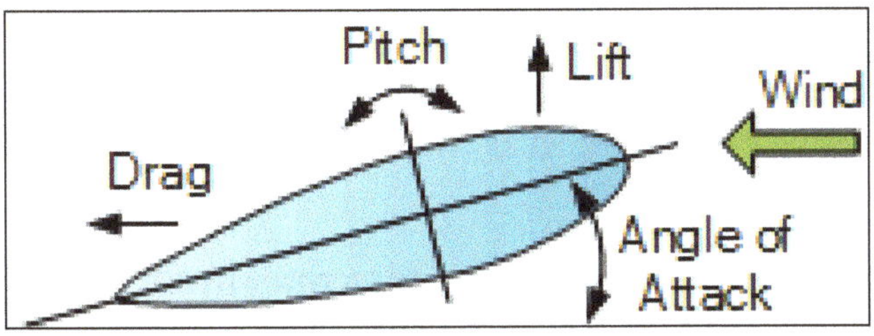

Therefore, to keep the rotor blade seeing an optimum angle of attack increasing lift and efficiency, wind turbine design blades are generally twisted throughout the length of the blade. In addition, this twist in the wind turbine design keeps the rotor blades from spinning too fast in high wind speeds.

However, for very large-scale wind turbine designs used for electrical power generation, this twisting of the blades can make their construction very complicated and expensive, so some other form of aerodynamic control is used to keep the blades angle of attack perfectly aligned with the wind direction.

The aerodynamic power produced by the wind turbine can be controlled by adjusting the pitch angle of the wind turbine in relationship to the angle of attack of the wind as each blade is rotated about its longitudinal axis. Then rotor blades with pitch control can be flatter and more straight but generally these large blades have a similar twist in their geometry but much smaller to optimise the tangential loading on the rotor blade.

Each rotor blade has a rotational twist mechanism, either passive or dynamic built into the root of the blade, producing a uniform incremental pitch control along its length (constant twist). The amount of pitch required is only a few degrees as small changes in the pitch angle can have a dramatic effect on the power output that the energy contained in the wind is proportional to the cube of the wind speed.

One of the major advantages of rotor blade pitch control is the increase in the wind speed window. A positive pitch angle produces a large starting torque as the rotor begins to turn decreasing its cut-in wind speed. Likewise, in high wind speeds when the rotors maximum speed limit is reached, the pitch can be controlled to keep the rotors rpm from exceeding its limit by reducing their efficiency and angle of attack.

Power regulation of a wind turbine can be achieved by using pitch control on the rotor blades to either reduce or increase the lift force on the blades by controlling the angle of attack. Smaller rotor blades achieve this by incorporating a small twist in their design. Larger commercial wind turbines use pitch control either passive, with the aid of centrifugal springs and levers (similar to helicopter rotors) or by active using small electrical motors built into the blades hub to rotate it the required few degrees. The principal disadvantages of pitch control are reliability and cost.

Blade Construction: The kinetic energy extracted from the wind is influenced by the geometry of the rotor blades and determining the aerodynamically optimum blade shape and design is important. But as well as the aerodynamic design of the rotor blade the structural design is equally important. The structural design consists of blade material selection and strength as the blades flex and bend by the winds energy while they rotate.

Obviously, the ideal constructional material for a rotor blade would combine the necessary structural properties of high strength to weight ratio, high fatigue life, stiffness, its natural vibration frequency and resistance to fatigue along with low cost and the ability to be easily formed into the desired aerofoil shape.

The rotor blades of smaller turbines used in residential applications that range in size from 100 watts and upwards are generally made of solid carved wood, wood laminates or wood veneer composites as well as aluminium or steel. Wooden rotor blades are strong, light weight, cheap, flexible and popular with most do-it-yourself wind turbine designs as they can be easily made. However, the low strength of wood laminates compared with other wood materials renders it unsuitable for blades with slender designs operating at high tip speeds.

Aluminium blades are also light weight, strong and easy to work with, but are more expensive, easily bent and suffer from metal fatigue. Likewise steel blades uses the cheapest material and can be formed and shaped into curved panels following the required aerofoil profile. However, it is much harder to introduce a twist into steel panels, and together with poor fatigue properties, meaning it rusts, means that steel is rarely used.

The rotor blades used for very large horizontal axis wind turbine design are made from reinforced plastic composites with the most common composites consisting of fibreglass/polyester resin, fibreglass/epoxy, fibreglass/polyester and carbon-fibre composites. Glass-fibre and carbon-fibre composites have a substantially higher compressive strength-to-weight ratio compared with the other materials. Also, fibreglass is lightweight, strong, inexpensive, has good fatigue characteristics and can be used in a variety of manufacturing processes.

The size, type and construction of the wind turbine you may need depends on your particular application and power requirements. Small wind turbine designs range in size from 20 watts to 50 kilowatts (kW) with smaller or "micro" (20- to 500-watt) turbines be used in residential locations for a variety of applications such electrical power generation for charging batteries and powering lights.

Wind energy is among the world's fastest-growing sources of renewable energy as it is a clean, widely distributed energy resource that is abundant, has zero fuel cost, emissions-free power generation technology. Most modern wind turbine generators available today are designed to be installed and used in residential type installations.

As a result, they are manufactured smaller and more lightweight allowing them to be quickly and easily mounted directly onto a roof or onto a short pole or tower. Installing a newer turbine generator as part of your home wind power system will allow you to reduce most of the higher costs of maintaining and installing a taller and more expensive turbine tower as you would have before in the past.

Wind Turbine Rotor Design

Wind turbine rotor blades are a high-technology product that must be produced at moderate cost for the resulting energy to be competitive in price. This means that the basic materials must provide a lot of long-term mechanical performance per unit cost and that they must be efficiently manufactured into their final form, including the cost of sufficient quality control. Unless a material choice and fabrication system can satisfy both of these requirements, it will not be appropriate for advancing the state of the art for economical production of power from the wind.

Both fiberglass-reinforced and wood/epoxy composites have been shown to have the combination of strength and low material and fabrication costs required for competitive blade manufacture. Their fabrication requirements and constraints, the current state of their materials database, and the areas where further research could advance their efficient use are, however, often quite different. The gradual evolution of wind turbine configuration and rotor designs also places a continuing demand on both systems to meet new requirements. In turn, the strengths and limitations of each material system also feed back into the process of system configuration selection and detailed design, as any design that cannot be efficiently produced will be at a competitive cost disadvantage. A really good design will already include the compromises necessary so that it can be effectively manufactured with acceptable quality in the material system chosen.

Airfoil Evolution

Early wind turbine blades typically used airfoil shapes borrowed from helicopter or low-speed aircraft use, such as the NACA 23xxx or 44xx series. These airfoils have shapes that are everywhere convex and in particular do not have a concave or reflex aft portion. While these were a natural starting point, they were found to be sensitive to the buildup of insects on the leading edge, which caused substantial loss of power output for many wind turbine designs.

Further evolution of airfoil choice in view of field experience has led to the use of the more modern Wortmann or NACA laminar designs, as well as the aft loaded LS-1 type. These airfoil designs generally employ a reflex or concave aft portion, which feature is also evident on the special-purpose SERI wind turbine airfoils. That concavity effectively eliminates certain manufacturing processes, such as filament winding, from consideration for producing the outer airfoil shell. While it also forces some modification to the female mold fabrication technique, both fiberglass and wood/epoxy composites have successfully been produced in reflex shapes without significant cost penalty.

Precise control of airfoil geometry is quite important in providing blades with consistent aerodynamic properties. Small variations in outboard airfoil camber (±1/4 percent of chord) or twist (±1/4°) can lead to substantial aerodynamic imbalance and rotor and turbine life reduction. For stall-controlled machines, off design peak power can further reduce cost-effectiveness by either overstressing driveline and generator components, and thereby increasing replacement costs, or by reducing total power output and revenue.

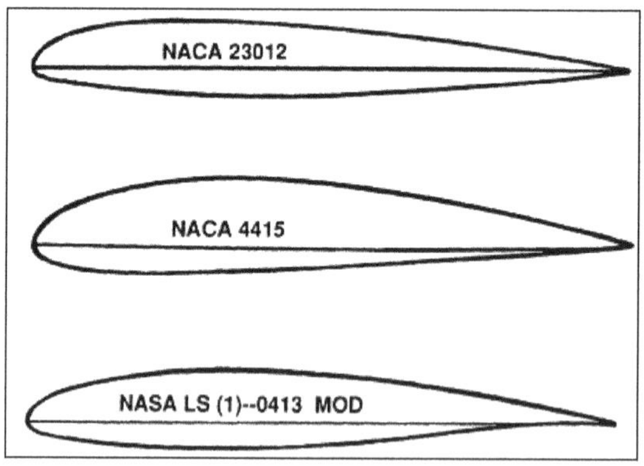

Commonly used early HAWT airfoils.

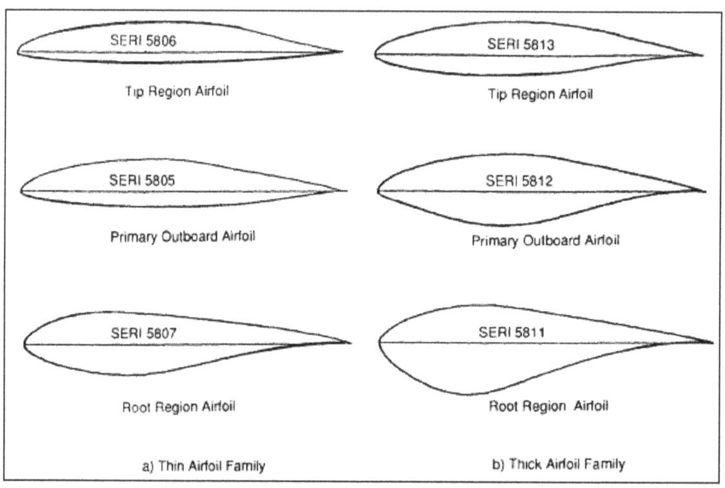

SERI advanced wind turbine airfoils.

This need for aerodynamic consistency and accuracy has led to the adoption of molding as the fabrication method of choice for both fiberglass and wood/epoxy composites, as it provides control right at the outer aerodynamic surface, which determines ultimate performance. Both material systems are able to provide the complete range of outboard airfoil shapes currently of interest.

While the outboard portion of the rotor changes little with material choice, the inboard region is a different matter. Fiberglass rotor blades often incorporate a large amount of inboard planform area and twist, and may carry the maximum chord quite far in toward the root. For a given rotor diameter, this will produce the most power, albeit at the cost of a significant increase in total blade surface area and materials. The flat sheet nature of the veneers used in wood/epoxy construction does not lend itself well to large inboard planform and the twist and the double curvature surfaces that result. Instead, a gradual transition from the inboard airfoil shape to an oval root is performed over the inner third of the blade. To regain the minor power loss due to lessened inboard planform area, a slight increase in blade length is provided. Since the turbine rotor designer is free to sweep energy out of the flow at whatever radius provides the least rotor cost, this is an effective solution because the cost of the slight extra length is small compared to the large reduction in inboard planform area. Low wind start-up torque is reduced, which could be a limitation for some turbine designs, but the reduction of planform area also reduces storm wind loading on the turbine as a whole, so the cost trade-off at the system level may or may not be favorable depending on start-up requirements.

Aerodynamic Tip Brakes

Turbine designs, which incorporate full span blade pitch control, can use that system to provide aerodynamic shutdown. Many current turbine designs, which use a mechanical brake for normal shutdown, do not have full span pitch control and must therefore depend on some other method of aerodynamic braking for safety. A great many methods have been tried, but most machines have used either a pivoting outboard tip section or a rotating tip plate.

The pivoting tip design has been dominant for the Danish fiberglass blades. The usual arrangement employs a centrifugal latch that releases the tip when an overspeed condition occurs. The tip is then allowed to move outboard, and a cam rotates it (in pitch) until it is perpendicular to the plane of rotation, thereby providing the drag needed to slow the machine aerodynamically. The typical arrangement to accomplish this uses steel parts for the mechanism, including cam, follower, attachment feet, and the structural tubes, which carry the tip loads back into the inboard blade. The feet that form the attachment to the blade are typically glassed (i.e., embedded in resin/fiber) onto one of the blade shells to secure the mechanism to the inner blade and to secure the tip to the pivoting tube.

Tip failures have been observed with this system, typically near the cut that separates the tip from the inner blade. That is where the bending moment in the tube is at the maximum, and indenting or scuffing from a loosely fitting bearing or collar can help initiate a failure. Damage in this area due to saw cuts, welding, or other manufacturing operations has also been known to lead to tube failure. These problems are not generally due to insufficient fatigue knowledge of the steels used, but rather to manufacturing and quality control issues, and the fact that a small geometric cross section for the embedded tube leads to rather high working stresses. The bonded retention of the assembly to the blade has generally been quite satisfactory.

Composites can be considered as a replacement for steel in this high-stress fatigue application. Fiberglass, while strong enough, has a low modulus that leads to large angular deflections and consequent problems with bearing alignment. Carbon fiber has the stiffness, strength, and fatigue endurance needed for the tip tube application, but it is considerably more costly than steel. However, as the cost of carbon fiber drops, the weight.

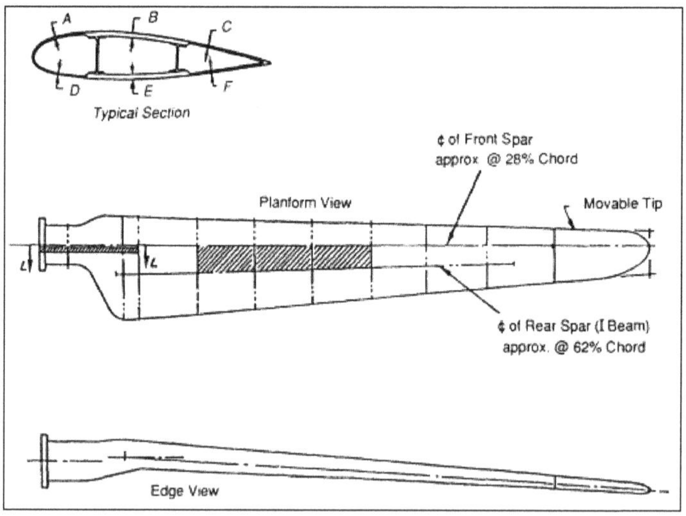

Nine-meter GRP wind turbine blade.

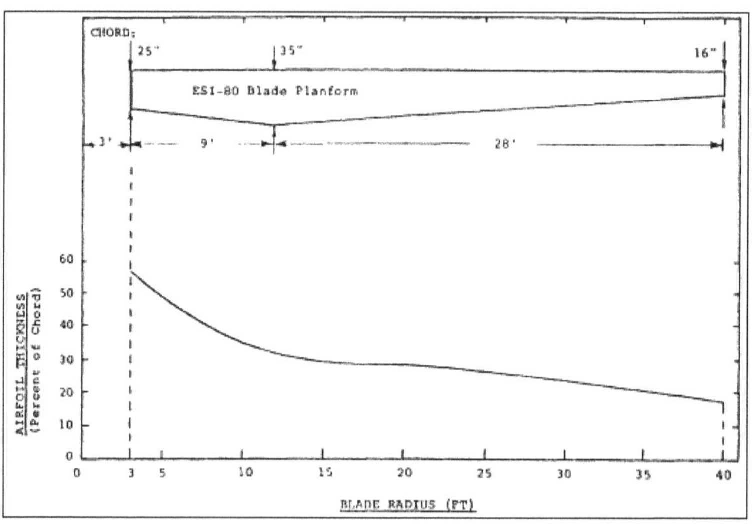

Eleven-meter wood/epoxy blade airfoil thickness distribution and planform.

Savings and corrosion resistance that it provides may make it attractive for this application.

The wood/epoxy composite blades produced by Howden also use the pivoting tip arrangement. While operating hours and machine numbers are not nearly as large as for the Danish designs, application with the wood/epoxy material system has not shown any unusual problems to date.

The Enertech and ESI designs of U.S. origin both used pivoting tip plates for overspeed protection. These were retained in place using threaded rod epoxy bonded into a veneer buildup at the blade tip. The load takeoff principle is identical to the well-tested root stud method and has proven highly reliable in service, except for cases of massive overspeed due to failure of the latch mechanism to deploy. This is not a material knowledge or fatigue problem.

Future wind turbines may employ techniques such as outboard blade ailerons or boundary layer control for overspeed protection. Such methods are not only potentially preferable from an aerodynamic standpoint, but also require much less disruption of the primary load path within the blade and could thereby reduce the possibility of long-term fatigue failure as well.

Blade Root Retention

Glass-Reinforced Plastic (Grp) Blade Roots

For both fiberglass and wood/epoxy composites, the blade-to-hub interface requires a high-performance fatigue design that can take loads from the composite blade structure into a metal hub assembly. Mating these dissimilar materials in high-cycle fatigue has been a serious design challenge in both material systems.

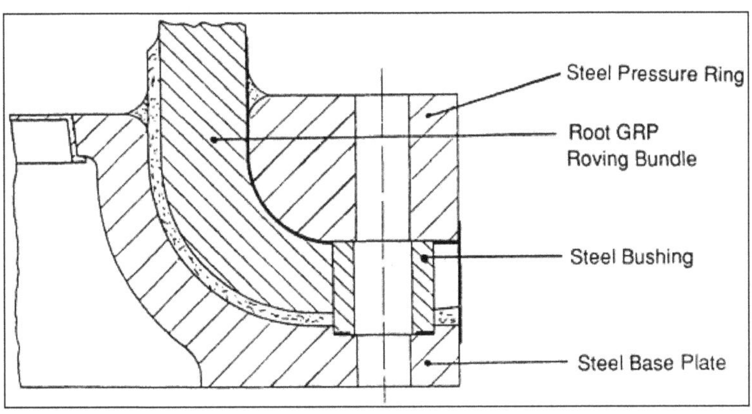

Flanged GRP blade root design.

Two major Danish fiberglass blade builders used a flanged root design. One variation, called the Hutter root after its originator, used thick unidirectional roving bundles wrapped around tubular bushings within the flange. These thick bundles bent around the root radius and terminated well up inside the root tube, so that an extensive bonding area was provided to the skin and spar reinforcing.

Since the bolts attaching the blade to the machine passed through the flange bushings, the blade was Mechanically captured both via the bolts/bushings and the flange clamping plates. Polyester-based filler material was used to fill the gaps between the hub, metal clamping plates, and the GRP flange to render the assembly as stiff and free of movement as possible.

The other major variation of the flanged root used root roving material that was molded uniformly outward to form the flange, after which the flange drilled and metal bushings were added. While the extra security of a wrap around the bushings was no longer present, mechanical clamping between the hub and flange plates provided positive mechanical capture. While filler material was again used to resist movement and render the final assembly as solid as possible, like the Hutter root, this design again did not depend a metal-to-GRP bond. Given the modulus and strain values of these designs, some motion between metal and GRP appears virtually unavoidable in the long run.

Both of these flanged root designs encountered considerable trouble in service. Partly due to quality control and manufacturing problems, and partly due to the inevitable shrinkage of polyester upon cure, the GRP flange was often not clamped well between the hub and flange plates. Once some gap occurred, the blade root was free to move somewhat and could then begin to reduce the filler material to powder. At the same time, the GRP bundles would begin to move relative to the steel flange plates and would often floss deep grooves into their surfaces due to the repetitive motion. Loss of root stiffness and the appearance of fine powder at the root were the symptoms of this degeneration. It was estimated by Stoddard that this problem was observed on 5 to 10 percent of the blades in service and was expected on 30 to 50 percent of the blades, given operation and management attention. This amounts to over 20,000 blades.

A procedure called a root pack was developed to deal with these root problems. In essence, the flanges were removed, and fresh filler material was added to take up the gaps and restore a tight fit. In many instances the bushings were shortened or recesses were machined into the flange plates so that the bolt pretension could clamp directly against the GRP flange. Bushings with a carefully selected crushing stress were also employed in some of the later designs, so that the GRP and steel bushing could each share some of the bolt preload, which was the original design intent of all the flange designs. It should be pointed out that some known flanged blades have over 5 years of operation with no problems and no root pack.

There is no doubt that the root pack procedure extended the useful life of many blades in the field. However, the increased clamping stresses within the GRP may now be taking a toll, as long-term tensile failure right through the fiber bundles has now begun to appear. While a well-designed, well-manufactured flanged root may offer life of several years, it does not yet appear able to offer the desired decades of life. This is true even though the blades in question are generally less than 10 m in length. Life expectancy is only likely to get worse as larger blades with increased gravitational bending moments are designed.

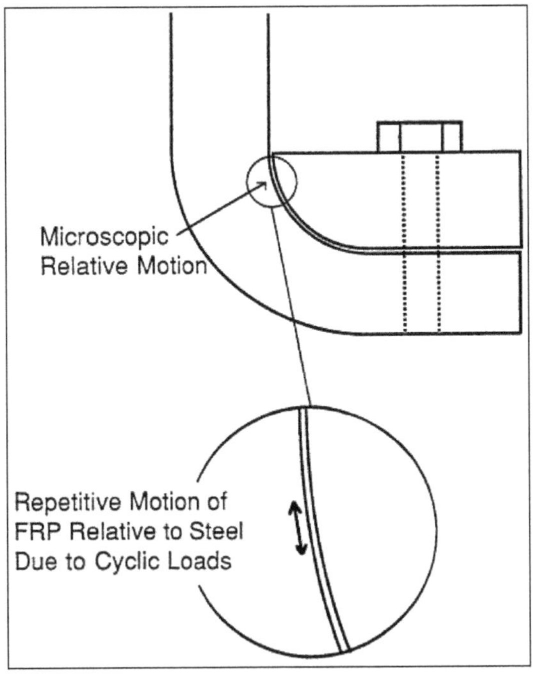

Flanged root design limitation due to steel/GRP strain incompatibility.

Another class of fiberglass root design is based on a tapered metal tube that is bonded to the root region of the blade. The tubes may be interior, exterior, or both and may occur with or without through bolts. For its long-term fatigue performance, this class of roots generally depends on a metal-to-GRP bond to transfer load via shear from one material to the other. These steel root tubes are provided with a steel flange so that they can be bolted to the hub assembly. The increased amount of metal required for this class of retention may be more costly compared to the types where the flange is formed directly from the GRP. However, roots of this sort appear to be reliable when properly designed and manufactured and have served well up to the present time.

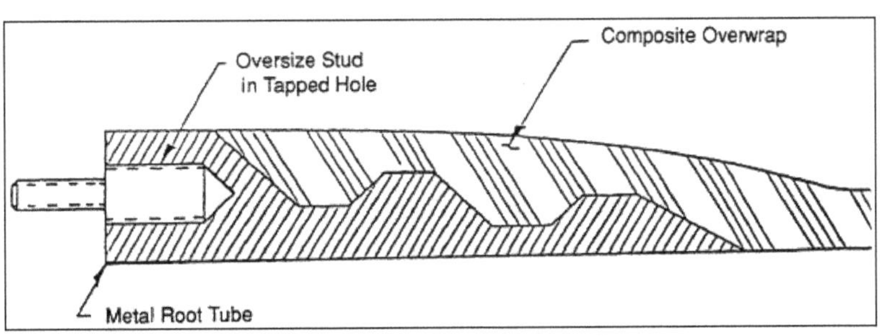

Bonded steel root tube GRP hub design.

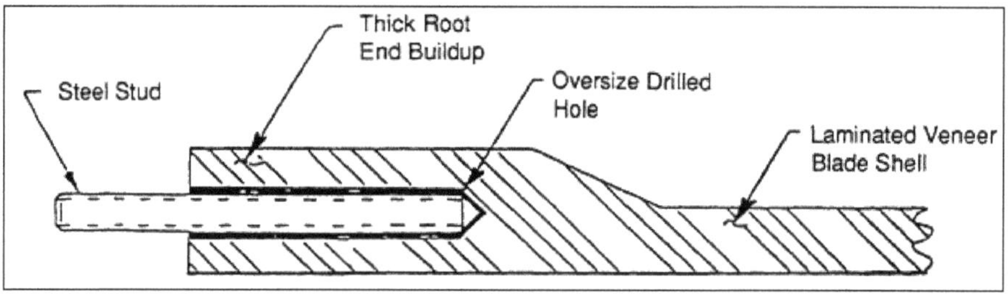

Wood/epoxy blade root.

Wood/Epoxy Blade Roots

The method of root retention for wood/epoxy composite blades has evolved along a quite different path than the flanged roots typical of GRP blades. While internal tubes with bolted flanges were briefly considered in the earliest days of the National Aeronautics and Space Administration's (NASA) MOD 0A program, they were quickly rejected as too costly and difficult. Attention was directed instead to the use of threaded steel studs, epoxy bonded directly into a thickened region of root laminate. Since the MOD 0A program required supporting a nearly 18.3-m (60-foot) blade on an 0.5-m (18.625-inch) bolt circle to match the existing hub, high mechanical performance was a necessity right from the start. While a simple bonded, threaded rod could not meet the one-time load and fatigue requirements, tapered designs were soon created that could meet the requirements with some margin to spare.

Once the difficult requirements of MOD 0A had been met, the small blade root designs for Enertech and ESI were not difficult, as a relatively larger and more optimally shaped oval bolt pattern could be specified at the time of blade design. This allowed a reduction in the number of studs, from the 24 of MOD 0A to only 10 or 12 for the smaller blades. Since the cost of the studs themselves is a substantial portion of the total root cost for a wood/epoxy blade, this reduction of part count was important from a cost standpoint. For the 20.4-m (67-foot) Westinghouse blades, a return to 24 studs was appropriate due to the much increased loads of that 600-kW, 43.3-m (142-foot) diameter machine.

The two primary manufacturing concerns for a stud root are that the epoxy completely fill the drilled hole without significant voids and that the studs be accurately coplanar so that stresses are not induced when mated to the wind turbine hub. Proper equipment and procedures have been developed to achieve these ends, and many fatigue tests of bonded studs, both singly and in whole roots, have shown that the system can readily tolerate the minor variations of a good manufacturing process.

Field experience with the Gougeon bonded stud root has been excellent. While some roots have failed in cases of machine runaway, these were at loads well beyond any intended design conditions. Fatigue failure has not been recorded for any of over 4000 blades in the field, with known use hours on some units at or beyond 25,000. Recent examination of a stud from the large-scale Westinghouse machine, with 17,000 hours, showed no evidence of fatigue damage. The extensive static and fatigue testing, which supported the early design efforts for this root retention type, apparently provided a solid basis for successful field application.

Other efforts to employ wood/epoxy-bonded studs have not been uniformly successful. The Howden 33-m turbine used this basic approach but is known to have had root problems. This choice of material system and root type is not enough to ensure success by itself. As in other parts of the wind turbine rotor, appropriate engineering design practice and fabrication technique are also required for success.

Theoretical analysis has indicated a small loss of performance when the bonded stud is used with GRP rather than wood/epoxy, but the performance levels are not reduced enough to prevent their successful application to that material system. Indeed, some of the Danish blades are now produced this way, rather than with the original flanged root designs.

Blade Joining

Efficient field joining of blade sections has not yet emerged as a major issue for blade design and manufacture, because most of today's machines are small enough that shipping the blades in one piece is not much of a problem. However, conceptual studies have indicated potential cost and weight advantages for two bladed rotors that are essentially one continuous structural piece due to the use of a composite "flow through" hub. This system entirely eliminates the normal root retention and, in Conjunction with a teeter system, needs to carry only net thrust and torque to the wind turbine.

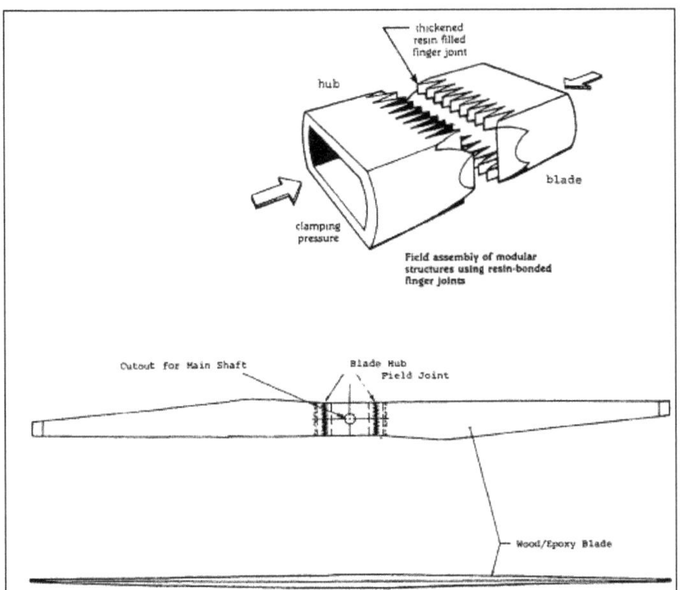

All-wood/epoxy composite Hub.

The resulting design is structurally efficient and should have excellent long-life fatigue potential, but it will require some form of field assembly to allow rotors to be shipped in pieces of convenient size. A cost-effective, fatigue-proven joining system for both fiberglass and wood/epoxy surely seems within reasonable technical reach but has not been developed and verified at this time. Such a field joining technique would also be of great value for vertical axis machines, whose curving blade shape makes them particularly difficult to ship in large pieces. Many other ways to use such a system to reduce machine-installed cost may be discovered once it is available and optimized.

Blade Design Considerations

Fiberglass Blades

The least-expensive structural material for fiberglass blade construction is unidirectional roving. This is a high value-added material for the manufacturer and, due to its straight fiber path and low resin content, is also one of the best-performing fiberglass forms on the basis of strength and stiffness to weight. It is therefore of economic and structural advantage to employ this material for as much of the primary structure as possible, consistent with efficient fabrication. Its limitations are that it does not have the thickness uniformity of fabrics and does not distribute well onto vertical or near-vertical mold surfaces. This can lead to fabrication difficulties near the blade root.

While conventional filament winding and transverse filament tape (TFT) are not applicable to the most recent classes of airfoils, with their generally reflex trailing edge regions, these fabrication techniques could still be considered for the primary structural spar in a spar/skin blade construction. It would be necessary for spar fabrication by winding to be very low cost to compete effectively with the unidirectional roving reinforcement mentioned above, but there may be designs or production levels for which that is possible. The method would gain considerably if some of the cost or complexity of root formation could be included efficiently in spar construction, as might be the case for winding directly over the root fitting.

While spar/shell construction has many attractions, it must overcome at least two significant structural handicaps. The first is that it is difficult for the spar to be stiff enough edgewise to prevent large strains in the blade trailing edge, which would lead to early shell cracking and failure. This could be handled by putting more active material in the blade shells so they carry more of the edgewise loads themselves or by the addition of a unidirectional trailing edge spline. The second handicap is that material concentrated in the spar rather than spread through the blade shell results in thinner skin laminates, which then need more buckling support and are easier to damage in the field. Simply put, it is difficult to beat the structural efficiency and ruggedness of a stressed skin design, and significant fabrication economies must exist for the spar/shell construction to be competitive with it. This is fundamental in assessing likely blade fabrication or material improvements.

The foregoing notwithstanding, it should be pointed out that several Danish blades, such as LM, MAT, and Vestas, are currently produced with spar/shell construction and are performing successfully. WEG is producing a wood/epoxy spar/GRP tail panel hybrid that also is performing well. Only time and further economic competition can show if such designs will prove to be the most cost-effective in the long run.

Even stressed-skin fiberglass blades often need panel stiffening in the relatively flat aft blade area, especially inboard where the span from shear web to trailing edge can be particularly large. Expanded foams are often used to provide the required panel support, due to their light weight, moderate cost, and easy formability. However, foams can be impact damaged by rough handling, and their long-term fatigue properties are not well characterized for the numbers of cycles typical of wind turbine blades, particularly under conditions of variable outdoor environment or in the presence of damage. Honeycomb is another skin-support possibility, but it has received little attention due to its cost and greater formability problems compared with other choices.

A panel-stiffening material used in a great many fiberglass blades to date is a felt-like material called core mat. Core mat is inexpensive and easy to conform to complex blade shapes. It is very light in the resin-free condition and, while it does absorb considerable resin, still has acceptably low density once bonded in place. Since this material is somewhat fibrous in nature, it may have rather good long-term fatigue properties, but at present this useful component of blade fabrication is not characterized in long-term fatigue. However, it can be said that shell-cracking problems seen in the field do not appear to be caused by core mat limitations for the kinds of lifetimes and cycle counts currently achieved. This situation could change for larger blades with larger panel sizes and deflections, but data for a quantitative assessment of these possible problems do not exist.

Since most blades are formed in female half-shell molds, joining of the blade shell halves and the interior shear webs is a major part of the overall manufacturing process. A mixture of polyester resin and talc called plasterite has been extensively used as a filler in the joining operations by certain fiberglass blade manufacturers. This material is stiff enough to attract considerable load when used in large quantity, as has often been the case, but is not possessed of fibrous reinforcement for strength and crack stopping in long-term fatigue. Another problem is shrinkage after fabrication, which builds in residual tensile stresses that the material is not well suited to handle. The only real attraction of this material appears to be its low cost, since the unfavorable combination of stiffness without strength appears to lead to high-cycle shell cracking, after crack origination in the bulk plasterite bonding material. Adequate fatigue characterization of this material does not exist, and it may be an impending problem for many blades now in use in the field. Other, better-suited, fillers with greater fatigue capability and suitable cost, handling, and health properties may be an important component in providing long-term blade life at a competitive cost. Test data to identify, select, and predict the performance of better fillers are not available to the blade industry at this time.

Another area of concern to the blade designer is the selection of blade skin-reinforcing materials. It is well known that conventional fabrics with a lot of fiber crimp due to the weaving process suffer in high-cycle fatigue due to resin breakdown and fiber stress concentrations at the intersections in the weave. So materials with less fiber crimp will be preferred in high-cycle fatigue wind turbine applications. However, this still leaves a wide selection of reinforcement styles across a broad range of costs in terms of the basic fiberglass chosen. A question of the role and durability of chopped mat and double bias constructions in the high-cycle environment also exists. Test data to allow a rational choice of the most cost-effective reinforcement are scant.

Another basic question exists in the choice of resin system. While epoxies are generally conceded to have the best long-term fatigue strength, they are considered by some to be more difficult to work with and too expensive for wind turbine blade fabrication. Current production has, therefore, concentrated primarily on the low-cost polyester system, with some designers opting for the somewhat higher-cost vinylester systems. While there is no doubt that vinylester is stronger at low- and moderate-cycle levels, it also appears to have a much larger fatigue effect than polyester. Some current data appear to suggest that, at 10 million cycles and above, polyester may, in fact, outperform the more expensive vinylester. An appropriate resolution of this question would obviously benefit cost-effective blade manufacture.

Wood/Epoxy Blades

The fundamental fatigue database for wood/epoxy composite is in good condition compared to that for GRP, as has been pointed out elsewhere. However, a number of topics to further improve the cost-effective use of this material system do exist. One of these, the cost-effective field joining of fabricated pieces for advanced rotor types, has already been mentioned. Additionally, research work by Gougeon Brothers completed under an SBIR contract to the Department of Energy has shown that advanced laminate splice joints can allow bulk laminate mechanical performance to approach that of unjointed material, but proving out practical manufacturing methods to realize this promise has not yet been accomplished. Another area of possible interest is that of alternative species to the Douglas fir used for the bulk of the previous work.

All of the above topics could improve the effectiveness of the wood/epoxy material system within the current design envelope. However, substantial gains may also be made by expanding the design envelope for the material. As was mentioned previously, the veneer sheets used with this construction technique constrain the acceptable blade geometries because they can accept only a very small level of double curvature without splitting or creating excessive resistance to the vacuum-molding forces. In this regard the problem is much like that for molding thermoplastic sheets, which are also relatively stiff and lack the convenient "drape" of conventional fiberglass materials. There may be many ways to move beyond current restrictions. For instance, high-temperature forming processes may be able to push the wood veneer into its plastic range and provide forming possibilities that room-temperature molding cannot. Or it may be possible to preprocess the veneer into a form that is much more conformable than the current full veneer sheets but that is still easy and efficient to handle. The fact that the database for wood/epoxy is in good condition for wind turbine blade fatigue design does not mean that further significant advances in this material system are not available, but rather that they may lie in somewhat different areas than for fiberglass.

Finally, both material systems can benefit from explicit feedback into the aerodynamic design process, so that the shapes specified exploit the strengths and cost efficiencies of each to the best overall effect.

Whichever material system is used, long-term fatigue performance will benefit from using simple external shapes with primary load paths that are as straight as possible, particularly in the highly loaded root and root transition areas. Minor power losses due to simplification of the inboard rotor geometry can be offset by a slight increase in rotor length, resulting in a potentially lighter and lower-cost blade of equal energy capture performance and superior long-term fatigue life. New blade designs should explicitly consider this structurally beneficial approach at the aerodynamic design phase.

Wind Turbine Blade Design

The reduction of fossil fuel dependency requires the expansion of the renewable energy sector. The European Union seeks to cover 20% of its energy needs from renewables by 2020. In order to achieve this goal, the wind energy capacity should be expanded by two orders of magnitude. The EU offshore wind energy capacity is expected to grow by 21% annually.

(a) (b)

Early history of wind turbines: (**a**) Failed blade of Smith wind turbine of 1941 (b) Gedser wind turbine (from).

The history of wind turbines for electric power generation started in 1988 Cleveland Ohio, USA, 1888 by Charles F. Brush and in Askov, Denmark in 1889 by pioneer Poul La Cour. In 1941, electricity production from wind was made using turbines with steel blades built by the company S. Morgan-Smith at Grandpa's Knob in Vermont in USA. One of the blades failed after only a few hundred hours of intermittent operation. Thus, the importance of the proper choice of materials and inherent limitations of metals as a wind blade material was demonstrated early in the history of wind energy development.

The next, quite successful example of the use of the wind turbine for energy generation is the so-called Gedser wind turbine, designed by Johannes Juul, with three composite blades built from steel spars, with aluminum shells supported by wooden ribs, installed at Gedser coast in Denmark in 1956–1957. After the 1970s, wind turbines were mainly produced with composite blades.

The Gedser turbine (three blades, 24 m rotor, 200 kW, Figure) was the first success story of wind energy, running for 11 years without maintenance. In this way, the linkage between the success of wind energy generation technology and the application of composite materials became an issue from the beginning: the first turbine, built with steel blades, failed, while the second one, with composite blades, worked for many years.

Composite Structures of Wind Turbines: Loads and Requirements

Composite materials are used typically in blades and nacelles of wind turbines. Generator, tower, etc. are manufactured from metals. Blades are the most important composite based part of a wind turbine, and the highest cost component of turbines.

A wind turbine blades consists of two faces (on the suction side and the pressure side), joined together and stiffened either by one or several integral (shear) webs linking the upper and lower parts of the blade shell or by a box beam (box spar with shell fairings). The flapwise load is caused by the wind pressure, and the edgewise load is caused by gravitational forces and torque load. The flapwise bending is resisted by the spar, internal webs or spar inside the blade, while the edges of the profile carry the edgewise bending. From the point of loads on materials, one of the main laminates in the main spar is subjected to cyclic tension-tension loads (pressure side) while the other (suction side) is subjected to cyclic compression-compression loads. The laminates at the leading and trailing edges that carry the bending moments associated with the gravitation loads are subjected to tension-compression loads. The aeroshells, which are made of sandwich structures, are primarily designed against elastic buckling. The different cyclic loading histories that exist at the various locations at the blades suggest that it could be advantageous to use different materials for different parts of the blade.

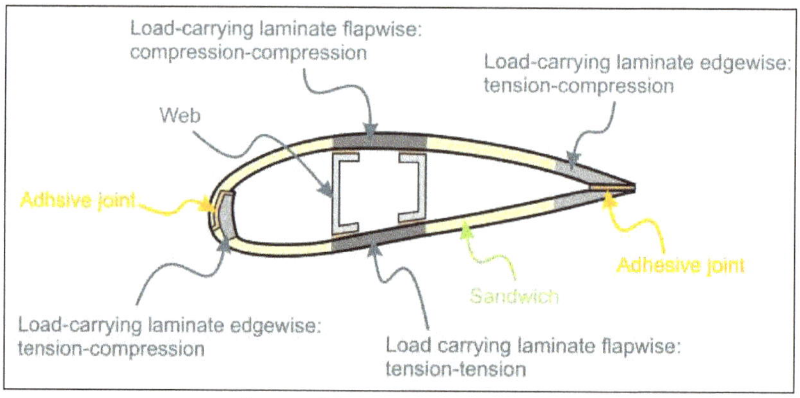

Schema of the section of the blade.

A major trend in wind turbine development is the increase in size and offshore placements. Increasing size is motivating by the desire to reduce of the leveraged cost of energy. With increasing size, the weight of the rotor blades increases, so that gravitational loads become design drivers. Also longer blades deflect more, so that structural stiffness (to ensure tip clearance, i.e., to avoid the blade to hit the tower) is of increasing importance. Thus, from a materials perspective, the stiffness-to-weight is of major importance. In addition, with the turbine designed to be in operation for 20–25 years, the high-cycle fatigue (exceeding 100 million load cycles) behavior of composites and material interfaces (bondlines, sandwich/composite interfaces) is of major importance.

Manufacturing of Wind Turbine Blades

During the first decades of the wind energy development, wind turbine blades were often produced using the wet hand lay-up technology, in open molds. The glass-fiber reinforcement was impregnated using paint brushes and rollers. The shells were adhesively bonded together/to the spars. This technology was used mainly to produce small and medium size blades (up to 35 and 55 m, respectively). For larger blades, the same technology was used, but the web were inserted and adhesively bonded between two sides, and the plies with more fiber content were used. The disadvantages of the open mold technology are high labor costs, relatively low quality of products and environmental problems. In 1970s, several companies and institutes explored the applicability of filament winding technology, seeking to improve the quality of turbine and to reduce labor costs.

The introduction of vacuum infusion and prepreg technologies allowed improving the quality of manufacturing. The prepreg technology, adapted from the aircraft industry, is based on utilizing "pre-impregnated" composite fibers, which already contain an amount of the matrix material bonding them together. Prepreg (widely used, for instance, by the Danish wind turbine producer Vestas) allows the industrial impregnation of fibers, and then forming the impregnated fibers to complex shapes.

The most widely used technology to produce the wind blades, especially longer blades, is the resin infusion technology. In the resin infusion technology, fibers are placed in closed and sealed mold, and resin is injected into the mold cavity under pressure. After the resin fills all the volume between fibers, the component is cured with heat. The resin infusion technologies can be divided into two groups: Resin Transfer Molding (RTM) (resin injection under pressure higher than atmospheric one) and Vacuum Assisted Resin Transfer Molding (VARTM) (or Vacuum Infusion Process) (when resin is injected under vacuum or pressure lower than atmospheric, typically, under a vacuum bag). A variation of VARTM called SCRIMP (i.e., Seemann Composite Resin Infusion Process) was developed in late 1980s and is quite efficient for producing large and thick parts. Currently, vacuum assisted resin transfer molding (VARTM) is the most common manufacturing method for manufacturing of wind turbine rotor blades. With his method, layers of fabrics of dry fibers, with nearly all unidirectional fibers, aligned in the direction along the length of the blade, are position on mold parts along with polymer foams or balsa wood for sandwich structures (for the aeroshells). In order to form a laminate that is thick by the root and gradually becomes thinner towards the tip, most plies run from the root only partly toward the tip; the termination of a ply is called ply-drop. The fabrics and subsequently covered by a vacuum bag and made air-tight. After the application of vacuum, low-viscosity resin flows in and wets the fibers. After infusion, the resin cures at room temperature. In most cases, wind turbine rotor blades are made in large parts, e.g., as two aeroshells with a load-carrying box (spar) or internal webs that are then bonded together. Sometimes, the composite structure is post cured at elevated temperature. In principle, this manufacturing method is well suited for upscaling, since the number of resin inlets and vacuum suction points can be increased. A challenge with upscaling is however, than quite many layer of dry fabrics must be kept in place and should not slip relative to each other. The composite is quite thick by the root section, typically exceeding 50–60 mm in the consolidated state. In practice, it can be a challenge to avoid the formation of wrinkles at double-curved areas and areas with un-wetted fibers and air bubbles can be entrapped in the bondlines. After manufacturing, the blades are subjected to quality control and manufacturing defects are repaired. Since a large blade represents a large value in materials, increasing sizes means that it becomes less and less attractive to discard blades with manufacturing defects. Thus, with increasing size the requirements towards materials go towards easier processing and materials should preferably be more damage tolerant so that larger manufacturing defects can be tolerated. Figure below shows the schematics of the manufacturing of a wind turbine rotor blade by assemblage and bonding of two aeroshells and two shear webs.

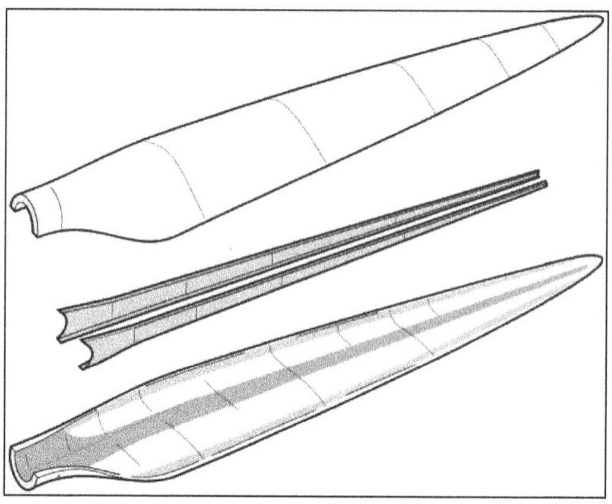

Schematics of the manufacturing of a wind turbine rotor blade by assemblage and bonding of two aeroshells and two shear webs (gray color indicates the primary load-carrying composites).

The infusion process is usually cheaper that the prepreg process. However, the prepreg composites have more stable, better and less variable mechanical properties than the composites produced by resin infusion. This technology is relatively environmental friendly, and makes it possible to achieve higher volume content of fibers, and to control the materials properties. Further, the prepreg technology allows higher level of automation and better choice of resins.

Lately, the automated tape lay-up, automated fiber placement, two-pieces or segment wind blades, enhanced finishing technologies are expected to come into use to improve quality and reduce costs of the composite blade manufacturing. A big challenge, in comparison with e.g., automatization of composite structures for aerospace, is the much larger thicknesses and the much larger amount of materials to be places in the molds for wind turbine rotor blades. For some parts of the blades, 3D woven composites represent a promising alternative to producing fiber reinforced laminates. Mohamed and Wetzel suggested producing spar caps from 3D woven carbon/glass hybrid composites. It was demonstrated that this technology allows producing spar caps with higher stiffness and lower weight, than the commonly used technologies.

Blade Damage

Precise information about the range and extent of damages found and repaired in operating wind farms is not generally available, however detailed studies of the composite material and adhesive interface damage found in wind turbine blades subjected to structural testing have existed for some time. The static loads and cyclic loading applied to the blade structure during full scale testing can result in the damage in the form of failure of various adhesive layers, laminate delamination, debonding at skin/core interfaces and splitting along fibers or in-plane compressive failure as well as gelcoat/skin debonding and cracks in the gelcoat. Obviously, damages in the primary load-carrying laminates (main spar and laminates at the leading and trailing edges) are of major concern. Fortunately composite materials are damage tolerant materials. Still, a major issue is that many of these damage modes are not easily detectable, since the damage do not originate from the external surfaces and may not be visible. For instance, in thick composite parts wrinkles may lead to the formation of compression failure and delamination. Cracks and delamination can also start from processing details such as ply-drops that locally causes a stress concentration. Cracks at e.g., trailing edge bondlines can be seen visually, but it is more difficult to assess how far they extend into the composite structure.

In addition to the various structural loading effects, wind turbine blades can also be subjected to lightning strikes, physical impacts and damaging surface erosion conditions whilst in operation. In certain rare but dramatic cases, a particular event can cause the total failure of a blade almost immediately; a powerful lightning strike or an extreme wind loading that leads to a rotating blade hitting the tower for example. Operators of wind farms take measures to minimize exposure of their structural assets to the full effect of storm conditions when these are forecast. But more commonly, over the course of a normal 25-year service life, it is expected that the composite material in a wind turbine blade will accumulate some signs of damage.

Blades are the most vulnerable parts of a wind turbine with respect to lightning. As every turbine can expect to experience a significant number of strikes during service life, all blades have a lightning protection system to reduce the effect of such strikes when they occur. Despite this it is common to observe scorching damage and cracking around the lightning attraction point of a blade as well as spar rupture, separation and surface tearing in more extreme cases.

A significant damage form observed in operating turbine blades is caused by (abrasive) airborne particulates impacting and eroding the leading edge, especially towards the tip where velocities are higher. Once established this rough surface will degrade the aerodynamic performance of the blade and reduce power production; if left unrepaired structural damage to the laminate material will soon develop requiring a longer and more complex repair effort.

Icing is the accumulation of ice on the surface of the blades under particular low temperature weather conditions. In extreme cases it will stop the operation of the turbine, but before that will disrupt the aerodynamics of the blade and reduce the energy generation as well as unbalancing the load distribution in the system and thus reducing structural fatigue lifetime.

Composites for Wind Turbine Blades

Fibers

Glass and carbon fibers: The stiffness of composites is determined by the stiffness of fibers and their volume content. Typically, E-glass fibers (i.e., borosilicate glass called "electric glass" or "E-glass" for its high electric resistance) are used as main reinforcement in the composites. With increasing the volume content of fibers in UD composites, the stiffness, tensile and compression strength increase proportionally, yet, at high volume content of fibers (after 65%), there might be dry areas without resin between fibers and the fatigue strength of the composite reduces. Typically, the glass/epoxy composites for wind blades contain up to 75 weight % glass.

Many investigations toward the development of fibers, which are stronger than the usual E-glass fibers, have been carried out. The high strength fibers (which are still used seldom in practice, but represent a promising source of the composite materials improvement) include glass fibers with modified compositions (S-glass, R-glass, etc.), carbon fibers, basalt and aramid fibers. S-glass (i.e., high strength glass, S means "Strength" here) developed in the 1960s, shows 40% higher tensile and flexural strengths, and 10–20% higher compressive strength and flexural modulus, as compared to E-glass. The S-glass is much more expensive than E-glass. S2 glass was developed in the 1968 as a commercial version of S-glass. S glass and S2 glass fibers have the same composition (magnesium alumino-silicate). The main differences are in sizing (fiber coating) and certification procedure. The price of S2-glass is around 10 times of that of E-glass. R-glass fibers, introduced in 1968, are produced with a calcium aluminosilicate glass with less silica and added oxides. Some other special glasses developed by Owens Corning are ECRGLAS, Advantex and most recently Wind Strand glass fibers. The

Wind Strand glass fibers show15 percent higher stiffness and up to 30 percent higher strength when compared to E-glass.

Carbon fibers are considered to be a very promising alternative to the glass fibers. They show much higher stiffness and lower density than the glass fibers, thus, allowing the thinner, stiffer and lighter blades. However, they have relatively low damage tolerance, compressive strength and ultimate strain, and are much more expensive than the E glass fibers. Carbon fiber reinforced composites are sensitive to the fiber misalignment and waviness: even small misalignments lead to the strong reduction of compressive and fatigue strength. Carbon fiber composites are used by the companies Vestas (Aarhus, Denmark) and Siemens Gamesa (Zamudio, Spain), often in structural spar caps of large blades.

Aramid and basalt fibers: Further, an interesting alternative is using non-glass, high strength fibers first of all, aramid and basalt fibers. Aramid (aromatic polyamide) fibers demonstrate high mechanical strength, and are tough and damage tolerant, but have low compressive strength, low adhesion to polymer resins, absorb moisture, and degrade due to the ultraviolet radiation.

Basalt fibers show good mechanical properties, are 30% stronger, 15–20% stiffer and 8–10% lighter than E-glass, and cheaper than the carbon fibers. The application of basalt fibers in small wind turbines have been demonstrated in, and the results were very encouraging. In the works, the basalt fibers were used as hybrids with carbon fibers.

Hybrid composites: Hybrid reinforcements (E-glass/carbon, E-glass/aramid, etc.) represent an interesting alternative to the pure glass or pure carbon reinforcements. Ong and Tsai demonstrated that the full replacement would lead to 80% weight savings, and cost increase by 150%, while a partial (30%) replacement would lead to only 90% cost increase and 50% weight reduction for 8 m turbine. The world currently longest wind turbine rotor blade, the 88.4 m long blade from LM Wind Power is made of carbon/glass hybrid composites.

In a number of works, the strength and damage mechanisms of hybrid composites were studied. It was reported, among others, that the incorporation of glass fibers in carbon fiber reinforced composites allows the improvement of their impact properties and tensile strain to failure of the carbon fibers. Manders and Bader observed an enhancement of the failure strain of the carbon fiber reinforced phase when “carbon fiber is combined with less-stiff higher-elongation glass fiber in a hybrid composite”. However, in, it was shown on the basis of computations that the dependency of the composite strength on the ratio glass/carbon is V-shaped, with a minimum at the content of the order of 60% carbon, i.e., the hybrid strength can be under some conditions be lower than the strength of both pure glass or pure carbon composites. This observation was confirmed experimentally in. Thus, while the hybrid composites seem to be a very promising group of composites for wind energy, additional investigations are required for the optimal composition of the materials. Figure shows a computational micromechanical model of hybrid glass/fiber composites, and the composite degradation process (first, failure of glass fibers and then failure of carbon fibers).

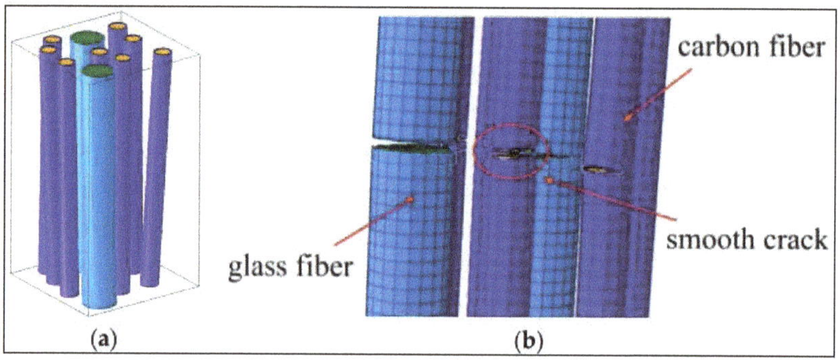

Micromechanical model of hybrid glass/fiber composites (a); and the simulated fiber failure (b). Reprinted from with kind permission from Elsevier.

Natural fibers: In some cases, natural fibers can be used as well. The advantages of natural fibers, as sisal, flax, hemp, jute, are the low costs, availability and environmental friendliness. The disadvantages are the quality variations, high moisture uptake and low thermal stability of the raw fibers. Holmes et al. tested a novel bamboo-poplar epoxy laminate for wind turbine blades, and demonstrated that this material has high strength and stiffness, and can be used in wind blades instead of common composites. The high strength and durability of bamboo as well as its quick growth and broad availability make the bamboo to a very promising material for the wind energy applications.

An interesting option for developing countries is small turbines, producible on-site, and made from "natural composites", i.e., from locally available timber. In a series of investigations, a group of Nepali, Danish and Australian scientists studied the applicability of different timbers for wind turbines, and demonstrated that the turbines with wooden blades represent a reliable and low cost option for the developing.

Testing the timber wind turbine blade (a) and small wind turbines installed in Nepal (b).

Matrix

Typically, thermosets (epoxies, polyesters, vinylesthers) or (more seldom) thermoplastics are used as matrices in wind blade composites.

Thermosets: Thermosets based composites represent around 80% of the market of reinforced polymers. The advantages of thermosets are the possibility of room or low temperature cure, and lower viscosity (which eases infusion and thus, allowing high processing speed). Initially, polyester resins were used for composite blades. With the development of large and extra-large wind turbines, epoxy resins replaced polyester and are now used most often as matrices of wind blade composites. Still, recent studies (e.g., by Swiss company DSM Composite Resins) support arguments for the return to unsaturated polyester resins, among them, faster cycle time and improved energy efficiency in the production, stating that the newly developed polyesters meet all the strength and durability requirements for large wind blades.

Further, the development of matrix materials which cure faster and at lower temperatures is an important research area.

Thermoplastics: Thermoplastics represent an interesting alternative to the thermoset matrices. The important advantage of thermoplastic composites is their recyclability. Their disadvantages are the necessity of high processing temperatures (causing the increased energy consumption and possibly influencing fiber properties) and, difficulties to manufacture large (over 2 m) and thick (over 5 mm) parts, due to the much higher viscosity. The melt viscosity of thermoplastic matrices is of the order 10^2–10^3 Pa s, while that for thermosetting matrix is around 0.1–10 Pa s. Thermoplastics (as differed from thermosets) have melting temperatures lower than their decomposition temperatures, and, thus, can be reshaped upon melting. While the fracture toughness of thermoplastics is higher than that of thermosets, fatigue behavior of thermoplastics is generally not as good as thermosets, both with carbon or glass fibers. Other advantages of thermoplastics include the larger elongation at fracture, possibility of automatic processing, and unlimited shell life of raw materials.

Nanoengineered polymers and composites: In several works, the possibilities of improvement of composites properties by adding nanoreinforcement in matrix were demonstrated. Additions of small amount (at the level of 0.5 weight %) of nanoreinforcement (carbon nanotubes or nanoclay) in the polymer matrix of composites, fiber sizing or interlaminar layers can allow to increase the fatigue resistance, shear or compressive strength as well as fracture toughness of the composites by 30–80%. Loos, Manas-Zloczower and colleagues developed various wind turbine blades with secondary carbon nanoparticles reinforcement (vinyl ester, thermoplasts, epoxy composites containing CNTs) and demonstrated that the incorporation of small amount of carbon nanotubes/CNT can increase the lifetime up to 1500%. Koratkar and colleagues studied graphene as a secondary reinforcement for the nanomodification of wind turbine composites, and showed experimentally that the graphene reinforcement is very promising in the development of stronger, long-life turbine blades for the wind industry. Merugula and colleagues estimated theoretically that the addition of 1–5 wt % of carbon nanofibers (CNF) to the interfaces of glass fiber reinforced epoxy composites for blades in 2 MW and 5 MW turbines leads to improved tensile stress and modulus, and allows

20% weight reduction of the blades, leading to the increased lifetime. One should note that transferring property improvements obtained in specific polymer-nanocomposites (without fiber reinforcement) as matrix material to laminates with reinforcing fibers remains an issue, especially with respect to volume fraction of the nano-fillers and the lower bound of the scatter in improvements obtained. In some cases, improvements from using nano-modified polymers as matrix (e.g., for improved strength or toughness) come with intrinsically lower property values in other areas (e.g., glass transition temperature) limiting the processability or the applicability of nano-modified polymers. In, the applicability of hierarchical composites for wind energy applications is analyzed, using the computational modelling. Also, the feasibility of using hybrid and nanoreinforced composites in wind blades, as a replacement for the currently used glass fiber/epoxy composites is evaluated in. It was demonstrated in numerical studies that the gains in the lifetime of the composites justify additional investments to produce the wind turbine blades from hybrid and nanoreinforced composites. Still, as noted in, there remains a lot of practical and economical challenges before the nanoengineered wind turbines are used. Figure shows a micrograph of a carbon fiber with CNT reinforcements in fiber-matrix interface (a) and computational model of a composite with secondary CNT particles.

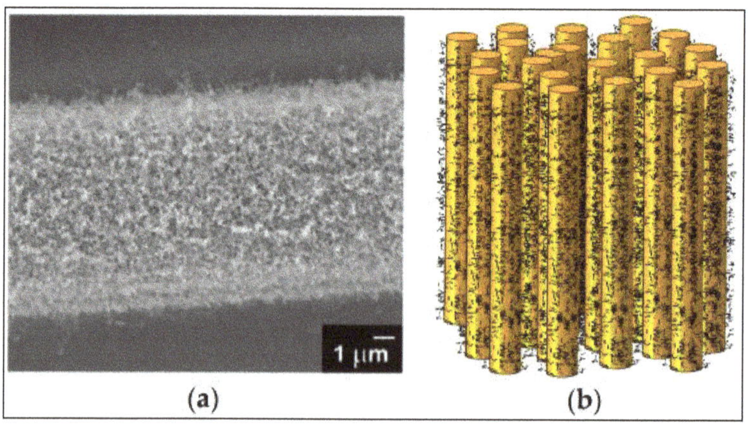

Micrograph of a carbon fiber with CNT reinforcements in fiber-matrix interface (a) and computational model of a composite with secondary CNT particles (b). (Reprinted from with kind permission from Elsevier).

Sizing

Fibers for composites are being applied with a surface sizing often referred to as sizing. The application happens during or just after the manufacturing. The two main reasons to apply sizing is to protect the fibers and to increase bonding to the matrix. The adhesion between fiber and matrix is pivotal when it comes to stress transfer between fiber and matrix and the mechanical properties of the composite. The stress transfer happens not only between the fiber and the matrix but across the sizing as well. Unfortunately the sizing and its reactions in the interface is barely understood due to the complexity and inaccessibility opposes thorough investigation.

Glass fiber sizing is applied as a part of the manufacturing process. After mixing and melting the glass components the melt is lead to a bushing with a large amount of holes from which the fibers are drawn continuously with an impressive speed of 2500 m/min. As soon as the melt leaves the bushing it is rapidly quenched by water being sprayed on the drawn fibers. The quenching ensures the amorphous network structure within the glass that yields the flexible. Several meters below the water spay the fibers pass a roller half immersed in sizing only close enough to pick up the liquid without touching the roller. The sizing contains around 3–10 wt % solid material in an aqueous suspension. After this step the fibers are gathered in a strand often referred to as a roving which contains in the range of 50–4000 single fibers. The fibers are then dried in large ovens above 100 °C for hours in order to evaporate surplus water and for curing of the sizing. Within seconds after the application the effect of the sizing is being put to work in regards to protecting the glass fiber surface against fiber-fiber damage in the rovings and in the later weaving.

The ideal sizing not only protects the otherwise fragile fibers during processing it also reduce fuzzy behavior, it disperse well on the fiber surface resulting in a homogeneous product, it ensure a good wetting during manufacturing of the composite yielding low a low void content, and it maximize the fiber matric interaction for optimum stress transfer. The multiple tasks cannot be covered by one compound thus the need of multiple components. The patents behind sizings have been studied and they reveal that sizings for glass fibers consist of minimum a film former and a coupling agent, but mostly more components are included. A large study indicated that the film former makes up around 80 wt % of the dry sizing and the coupling agent around 10 wt %. The task of the film former is to protect against fiber-fiber damage and to protect the roving during winding and weaving. It is often a polymer similar to the matrix that the end-product aims for e.g., polyesters, polyurethanes, and epoxies yielding a good wetting during composite manufacture. Reduction of the stress corrosion triggered by water is attained by the addition of a coupling agent.

This is often chosen to be an organosilane and in some cases chromium or titanium oxides. Organosilanes has the possibility to react with the glass fiber surface through a sol-gel reaction which can covalently bond the organosilane or a polymeric form of the organosilane to the fiber surface. With a functionality of the organosilane that complement the matrix it is possible for these to react thus forming a connection between the fibre and the matrix. This is the reason why the coupling agent is considered a crucial parameter in regards to the adhesion between fibre and matrix. The most used silanes have amine or epoxide functionalities. Anti-static agents reduce the fuzziness of the fibres and thereby helps form the roving. Emulsifier agents stabilize the insoluble components that are added to the sizing suspension. Furthermore they reduce the formation of foam and adjust the viscosity of the sizing. Lubricants improve the dispersion on the glass fibres and help protect the surface. Acid or alkalis can be added to adjust the pH to around 4 in order to facilitate the hydrolysis of the silanes. Wetting agents and anti-oxidants can also be added to sizings.

Despite that carbon fibers are organic and therefore don't have the same compatibility issues as glass fibers they still need sizing to manage the smooth and often inert surface. Carbon fibers receive both a surface treatment and a sizing both of which are conducted after the manufacturing of the fibers. The manufacturing of carbon fibers is much different from glass fibers. The steps include polymerization from a precursor, stabilization, and carbonization followed by surface treatment and sizing application. The surface treatment etches or roughens the surface using an anodic or nonionic electrolytic polymer e.g., epoxy, polyamide, polypropylene, or polyurethane dispersions. The result is an increase of the surface area which then increases the bonding between fiber and matrix. An increase in the fiber roughness is also expected to increase the interfacial sliding friction between fiber and matrix after debonding. The amount of sizing is in the range of 0.5–5 wt %. As with the glass fiber sizing it does more than protect the fibers it also facilitate strand formation, reduce fuzz, improve processability and increase bonding between fiber and matrix. The sizing is often an emulsion of polymeric components.

Natural fibers are hydrophilic and bond poorly with both thermoplastic and thermoset polymers. Very little work has been done in the field of sizings for natural fibers and using glass fiber sizings is not the best solution. As a coupling agent it is mostly silanes and isocyanate that are selected. Physical treatments as plasma or heating are also used to alter the fiber surface for better bonding to the resin. The treatment is carried out as a part of the processing before being woven. Sometimes the coupling agent is an additive to the matrix resin instead of a surface treatment of the fibers, making it less efficient in composites with low fiber content as modified polymers can hide within resin rich areas with little contact to fibers.

Testing, Degradation and Computational Modelling

Testing of Wind Turbine Blade Materials and Structures

In the design process of wind turbine blades, tests on several scales can be performed in order to measure the relevant material properties and to check the accuracy of the computational design models used to estimate the load bearing capacity. However, currently only coupon and full-scale tests are required in the IEC 61400 standard for wind turbines in order to certify wind turbine blades.

At coupon level, small test specimens with the basic material are tested in order to determine the material properties and their statistical characteristics in both ultimate and fatigue limit states. The test specimens at coupon level are normally relatively inexpensive to produce (small amount of materials), are normally relatively easy to model and interpret and with several different tests with many repetitions a good description of variation can be obtained.

At subcomponent level parts of a wind turbine blade are tested in order to determine selected parts load bearing capacity and verify computational models for potential critical details. Subcomponent tests are in general more expensive and complicated to test than coupons for which reason fewer tests are performed with each subcomponent.

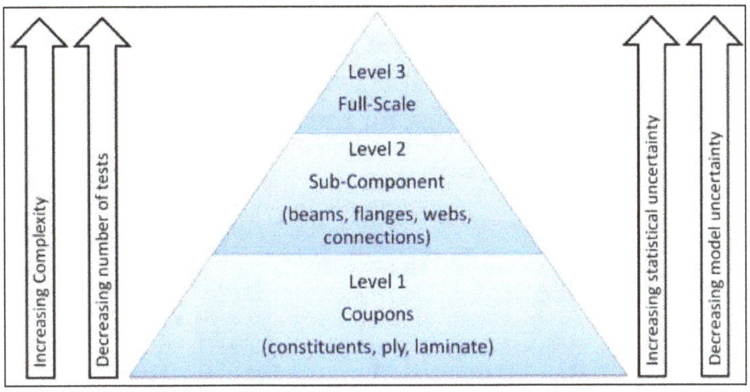

Illustration of type of tests, which can be performed for assessment of load bearing capacity of wind turbine blades.

At full-scale level prototypes of the blade are tested both dynamically and statically following the requirements in the IEC 61400-23 standard on full-scale testing. Full-scale blade tests are performed on typically one or two blades in order to verify that the blade type has the load carrying capability and service life provided for in the design. Since the cost of a blade itself is high, the blade is large and usually equipped with a lot of transducers, sensors and instruments, and the time needed for the dynamic test and the subsequent data analysis can be several months for large blades, the cost due to waiting time for market introduction is also significant.

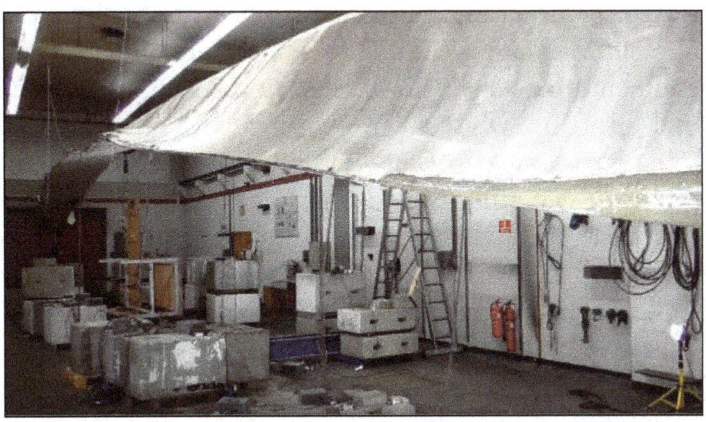

Full-scale testing: A 34 m long wind turbine blade subjected to static test in a combined flapwise and edgewise load direction.

Initial work on how to plan and apply subcomponent tests in the design process of wind turbine blades were done in the project: "Experimental Blade Research—Phase 2 (EBR2)". Parts of this work was then used for making the new DNV GL rotor blade standard DNVGL-ST-0376, which for the first time makes it possible to use subcomponent testing as part of blade certification. A subcomponent test method designed to check the compressive strength of the trailing edge region in wind turbine blades under a simplified loading were first proposed by DTU and then further developed under the EU-funded project IRPWind in corporation with Knowledge Centre WMC and Fraunhofer IWES. In finite element simulations show that the proposed static subcomponent test method is promising in obtaining a test of the compressive strength of the trailing edge region under a simplified loading. It is overall found that the failure load and failure mode is very similar to full blade test for the analyzed test specimen.

Sub-component testing of trailing edge panel in compression using the method developed at DTU.

Currently the IEC61400-23 and DNVGL-ST-0376 standard require that blade fatigue testing is carried out by testing blades in two directions—flapwise and edgewise—one direction at a time. During their lifetime wind turbine rotor blades are exposed to highly dynamic loads, resulting from cyclic changes in gravity direction, centrifugal forces, and changing wind conditions such as average wind speed, turbulence intensity, rapidly changing wind direction, wind shear, extreme wind gusts and site-specific loads like e.g., wake effects from neighboring wind turbines. The broad and complex load spectrum results in the accumulation of a significant amount of fatigue damage over the turbine lifetime. Fatigue of materials and interfaces are therefore major failure mechanisms in wind turbine blades. The currently required fatigue testing methods are not representing the real service loads very well and there are therefore attempts to develop more realistic test methods. One method is dual-axis fatigue testing where the flapwise and edgewise directions are tested simultaneously. This approach is shown in to be more representative of the loading seen in service and can thus contribute to a potentially more realistic testing of wind turbine blades.

Mechanical Degradation of Wind Blade Composites

The in-plane material degradation of the wind blade composites under loading can start at the microscale cracking, e.g., by fiber failure. Once a fiber fails, a local load redistribution occurs. An area with a broken fiber will induce a higher stress concentration in the matrix, which can lead to the matrix cracking. If the fiber/matrix interface bonding is sufficiently weak, cracks form on interfaces (fiber/matrix debonding), leading to the fiber slip along the fiber/matrix interface. This can potentially promote further matrix cracking and debond crack growth. For delamination cracks, if the crack bridging mechanism is operative, the load is shared by the bridging fibers and crack tip, and the stress intensity factor on the crack tip is reduced. A higher amount of bringing fibers leads to the lower stress intensity factor on the crack tip, and the resistance to crack growth increases with increasing the crack length (R-curve behavior). The extension of a crack, bridged by intact fibers, leads to the debonding and pull out of fibers that increase the fracture toughness of the material. The fiber debonding, fiber fracture and pull-out are the most important toughening mechanisms in fiber reinforced composites.

In the regions under compressive loading (the downwind side of the blade and spar) fiber crushing and shear banding can be observed. In, the damage mechanisms of glass fiber composites under compressive loading was studied experimentally, using SEM observations, as well as numerically. The damage mechanisms under cyclic loading (fatigue) are in many cases different from the static damage mechanisms. For multidirectional laminates, the longitudinal plies (with fibers aligned in the direction of tensile load) control the fatigue behavior and lifetime of the composites. The presence of backing fibers (i.e., fibers oriented off-axis to the load direction) can also have a negative effect on the fatigue life of the composites.

Under tensile cyclic loading along the fibers, damage mechanisms are similar to those in static loading when high tensile strain is applied (so-called "Region I" of the fatigue-life diagram, corresponding to high loads and low lifetimes). If small strains are applied, the damage growth is rather slow or even negligible, and does not lead to the material failure even after 108–109 cycles of loading. Under the cyclic loading which is between these two regions, progressive fiber bridged cracking, progressive debond propagation and stochastic fiber breakage are observed in the composites. In laminates with off-axis or backing plies, the damage often begins in the plies oriented transverse to the primary loading direction. In these plies, a crack forms from coalescence of fiber-matrix debonds. When the crack forms in transverse plies, it can lead to the stress concentration in nearby off-axis plies.

Computational Modeling of Wind Turbine Materials

The aerodynamically optimal shape of wind blades corresponds to the much lower blade thickness than that dictated by the structural design requirements. However, the structural design of blades is carried out on the basis of safety factors, which should take into account the unknown effects in the loading, degradation, cracking and failure of blades. It is often assumed that the safety factors used in wind blade design are taken too high. In order to determine safety factors more precisely, detailed information about the effects of different loading conditions (extreme winds, moisture, temperature variations) and the microstructure parameters of wind blade composites on their strength and lifetime is required. Further important tasks of the analysis of wind blade materials are the prediction of strength and the lifetime of the wind turbine rotor blades, analysis of the effect of different service conditions on the blade performances and analysis of reserves of improvement of the reliability and performances of wind turbines related to the microstructures of the wind blade materials (such as fiber sizing, clustering, nanoscale-engineered matrix, etc.).

This information can be obtained from computational models and theoretical and numerical studies of the wind blade material behavior under loading. For the modelling of damage processes, a series of analytical and numerical methods are used. Among the analytical methods used, one can mention shear lag-based models, fiber bundle model and its generalizations, fracture mechanics based and continuum damage mechanics based models. In the shear lag model, the force balance in a single fiber is considered assuming that the load transfer from fiber to matrix occurs only via shear stresses. In the fiber bundle model, the damage evolution in a bundle of fibers after one or several fibers fail is considered using statistical models of fiber strength and different load sharing rules. Using the shear lag and fiber bundle models, one can simulate the damage evolution in composites. In the continuum damage mechanics, the influence of many microcracks on the material behavior is modeled as tensor characterizing the reduction of the effective areas of a material due to the microcrack growth. In the framework of analytical micromechanics of materials, the relationships between volume content of reinforcements, its shape and orientation and the mechanical properties of materials are established using the variational principles of mechanics of materials and other continuum mechanics methods. These relationships allow studying the effect of the material structure, reinforcement geometry and distribution on the elastic properties of the composites. For the analysis of microstructure-properties relationships of wind blade materials, taking into account complex loading, initial and evolving microstructures (e.g., damage), numerical micromechanical methods are applied. Most often, the problems are solved using the finite element method, in which the solution of integral and differential equations describing the material deformation and microstructure evolution are obtained on the basis of discretization of the considered bodies and discrete approximations of the equations. Multiscale models of wind turbine blades, including also the material degradation at several scale levels, represent an important direction of the wind blade materials analysis.

Damage in Operating Wind Turbine Blades: Inspection and Monitoring Tools

Minor damage in the composite can be tolerated if it does not impair the structural performance of the turbine or risk propagating under normal operating conditions. But some forms of damage, once present in the structure, will propagate quickly and reduce the performance of the turbine or even cause an instability that will overload other structural components and potentially cause structural failure of the turbine. Clearly this will halt the operation of the turbine. Once this has occurred the wind farm operator will face a significant cost to make the necessary repairs before the turbine can be put back into operation. Although no strict guidelines exist to determine the criticality of different damages and defects to be found in operating WTBs, all wind farms have an inspection and maintenance procedure for their blades. The justification is that by checking the condition of the structure regularly, it is possible to schedule a series of minor repair tasks that will reduce the financial risk of an unscheduled major repair becoming necessary.

This process requires firstly an inspection of the current condition of all the blades. Previously requiring direct access by maintenance crew using climbing ropes or a crane lift, this task is more commonly done now with high resolution camera images taken from the ground an inspection platform, or mounted on drones. These images can be handled by computer software that "stiches" all the images together to allow a detailed overview of the entire surface area. Image analysis software can then compare previous inspections to the most recent and highlight any "exceptions" that should be checked by a trained blade engineer. In this way surface cracks (and also occasionally surface dirt and lighting/shading effects) will be registered for more detailed inspection in the follow-up phase.

Many severe damages are sub-surfaces and it is not always possible to make a judgement on the severity of the damages via an image. Therefore, a close inspection by a maintenance engineer must also be scheduled. Visiting all the areas highlighted by the initial inspection, the engineer will gather more information on each damage area (perhaps also discovering new ones) and generate a recommendation for each blade's repair requirement. The inspection is commonly based on the simplest and most robust techniques of visual assessment and manual tap testing. From this effort a job list (and a cost estimate) for the entire wind farm repair will be produced and approved.

It is common for external contractors to bid for the maintenance contract on large wind farms. Although lucrative, it is a competitive business and as the wind industry has grown and matured, access technologies and logistic advances have, together with the competition for contracts, had a lowering effect on the market price. Resisting this is the trend for larger, more complex and damage critical blades being placed in remote offshore wind farms. So maintenance and repair for blades in wind farms is still a significant portion of operating costs. And for operators there is also the consideration of risk around ensuring trustworthy third party inspections and repairs. On-site repairs are routinely documented with an annotated photograph of the pre- and post-repair damage. More analytical evaluation of the repair effectiveness is not conducted.

The idea of introducing a degree of automation into the entire inspection process for wind turbine blades has been investigated for some time. Sensors mounted in or on the blades provide continuous data remotely to the wind farm operator that can then be used to make best use of the available maintenance manpower. But implementing this in operating wind farms would incur additional expense at the highly price-focused manufacturing stage, where embedded sensors would need to be integrated with the structure. And transporting expensive and fragile measurement and inspection equipment to the harsh operating environment of offshore wind farms has similarly limited the uptake of more advanced NDT hardware. Whereas the robust nature of visual/ manual inspection (now streamlined and improved by tele- or drone photography and image analysis software) has been favored.

Advanced inspection tools (such as Ultrasonic scanning and thermography) are routinely used in the industry to provide Quality Assurance and Control from blade manufacturing. And full scale blade test facilities use sensor technology (resistance strain gauges, fiber optics, acoustic emission, digital image correlation, etc.) to provide real-time information about the response of the structure to various load conditions, and warnings about the occurrence and severity of any damage events. So there is no question that there exists inspection and monitoring technology that can be applied. The challenge that has yet to be solved is the relevant integration of robust, low-cost monitoring and inspection technology into the entire lifecycle analysis of the large wind farms that provides incontrovertible evidence of an improved exploitation of the structural asset. The discussion on detection tools is a balance between sensors with high sensitivity (enabling very early detection of small damages) possible requiring many sensors with small spatial coverage to sensors with less sensitivity, and thus much later detection of the damage, but requiring less sensors.

The offshore wind sector will continue to expand over the next several years (under 20 GW in 2016 to over 150 GW in 2030); in parallel to this, sensor (and inspection) technology will continue to become cheaper, more robust, with a higher functionality, and more widely applied and integrated within industrial systems and processes. In addition, the materials and structural design and manufacturing (plus repair procedures) will also advance. It seems inevitable that some combination of sensorised multi-functionality will shortly be incorporated into blades that are specifically designed to take advantage of these new understandings emerging around damage detection and damage tolerance criteria.

Vibration based damage detection systems rely on analysis of the dynamic response of the blades, either during operation or following an applied mechanical input. Damage identification is commonly based on the comparison between an undamaged and a damaged state. Ideally the ambient energy generated by the turbine operation would be used as the excitation source, however under realistic conditions detecting damage with simple analysis is challenging as the same order of modal property variations are generated by environmental effects and noise contamination. Therefore, more sophisticated methods must be deployed to create a reliable SHM system, and this can limit general application. For vibration based damage detection, more success has been reported using an external shaker or embedded actuator as using these a well distributed excitation is created within the entire structure, and a flat spectrum is generated in the frequency range of interest. Vibration based techniques have the advantage of being a mature, well-proven and cheap solution with respect to wind turbine gearbox and bearing fault detection systems. Despite the challenges involved in monitoring the more complex blade structure, some successes have been demonstrated. However, structural response changes will only detect relatively large damages in the blades and this limits the usefulness of the technique.

Fiber optics embedded in the blade structure can be used to measure strain. Several fibre Bragg gratings (FBGs) can be multiplexed within a single measurement fibre optic and with high sensitivity and reliability return local point strain measurements. Combined with their long fatigue life and immunity to electromagnetism, fibre optics are a promising sensor technique for integration within fibre reinforced plastic structures. FBG based systems are the most technologically ready fibre optic measurement system currently with commercial systems reducing prices and the size of installation hardware, as well as improving their robustness. The problem of fragile, bulky and expensive fibre optic measurement hardware is still true for most other (non-FBG) forms of fibre optic systems. For example, distributed sensing based on optical backscatter reflectometry recognizes changes in density and composition along the entire length of an unmodified optical fibre. Once recognized by the system, this "fingerprint" value will change due to any local variations in strain and temperature allowing measurements to be returned all along the fibre with a high resolution. This gives far more detail than even a heavily multiplexed FBG array can provide. The technique however, is currently limited to static testing as the extremely low levels of natural back scatter means signal to noise ratio is unacceptable in a dynamic situation.

Acoustic emission (AE) involves detecting transient bursts of elastic stress wave energy released by the formation of damage within the structure. Usually achieved with surface-mounted piezoelectric sensors which transform the elastic energy into an electrical waveform which can be processed and analysed. Detectable AE signals occur when a fibre composite material begins to experience local fiber failure, debonding, matrix cracking, delamination and splitting as the structure is placed under load. FRP structures generate huge numbers of such AE signals when loaded above previously reached maximum loads, as well as at lower load levels when previously formed damage is already present in the structure. This means that AE is a simple, useful and intuitive tool for detecting and locating damage during full scale testing in both static and dynamic loadings. The frequency range of useful AE signals is between 100 kHz and 1 MHz which above that used in vibration-based sensing (and in human hearing), but below the range designated as ultrasonic. Acoustic emission monitoring systems are used commercially in rotating machinery, metallic structures, bridge structures, and simple composite structures like pressure tanks. However, in-operation wind turbine blade monitoring is not a commercial activity due to the expensive hardware, the large data sets generated by the high sampling rate, and the attenuation rate for this frequency range in composite materials limiting the sensoric range meaning many sensors would be required to fully instrument each large wind turbine blade.

Acoustic emission is a passive technique that detects energy generated by the structure, in contrast guided wave technology (or acoustoultrasonics) is the terms used when the piezoelectric sensors are instead used as active transducers to generate a pre-defined input signal that can propagate through the structure and be detected by neighboring sensors. This known "Pitch-Catch" configuration will be altered when the signal interacts with damage or other factors that affect wave propagation between the input and detection points. In this way a network of sensor transducers can monitor any changes occurring over an entire structure. Commercially this technology is successful in simple oil and gas pipeline monitoring, however the complex composite material and dynamic structural environment of the wind turbine blade is proving a challenge. The use of guided wave technology for this application is therefore still in development, but is considered a promising technique.

Recycling

As pressure is put on to have greener and more sustainable products, the recycling of wind turbine blade have increasingly attracted the interest of wind turbine blade manufacturers and owners. However, recycling blades remains a challenge. The difficulties related to the process are mainly due to the structure of the blade and to the composite materials used.

Playground design.

The structure of the blade, as represented in Figure, is made of several elements, namely shear webs, load carrying beam, leading and trailing edge and aerodynamic shell. Depending on the blade manufacturer, the design and the arrangement of these elements will be different. In general each of these elements is consisting of a specific type of composite and the blade is manufactured as a one-piece component. To separate the different elements, the locations of the elements need to be known and a saw with diamond blade and sufficient water cooling is required. Due to that complex structure, it is difficult to recycle blades into any other application than blade. In addition, the blades to be recycled will be found in various conditions. Decommissioning of wind turbines can be decided as the turbines are reaching end of life, but also at earlier stage if it becomes interesting to replace the turbines by newer models or because the turbines were prematurely damaged. As a result, the quality of the material found in blades and the quality of the blade structure will be varying from blade to blade. The assessment of the blades conditions also represents a challenge. Visual inspection, which is normally used to determine the conditions of blades during inspection, does not reveal the presence of potential sub-surface damages. Finally, the amount of material coming from blades will fluctuate greatly as material will sporadically come from the decommissioning of single turbine or large windfarm. To summarize, the amount of material to be recycled coming from wind turbine blades will be varying in design and material, in quality and quantity. The development of a sustainable recycling solution for blades is therefore very complicated.

The other challenge in recycling blades is related to the composite material used in blades, which are made of a thermosetting matrix and glass fibers or a combination of glass and carbon fibers. Unlike thermoplastics, thermosetting matrix cannot be remolded to form new product. So the options are either to reuse the blade and the composite material elements as they are found in the blade or to transform the composite material into a new source of material. The first option only requires cutting the blade, while in the second option, heavier and more advanced processes need to be use. The first options will necessarily lead to a limited number of possible applications, while the second options will open up to many more. As an example, figure shows a playground made out of entire sections of blade. The structure of the blade is reused, but the application will be difficult to upscale to an industrial scale solution. Regarding the solutions involving heavier reprocessing, research on composite recycling has focused on processes to separate the fibers from the matrix to reuse them in new polymer composite applications. To do so, heat is necessary to degrade or dissolve the matrix material. The temperatures used in the recycling processes vary from 280 °C for a supercritical fluids process to 450 °C for a fluidized bed process. The issue is that the heat treatment will be detrimental to the mechanical properties of glass fibers, which will become extremely brittle. The glass fibers properties are however not the only challenge to overcome. The recovered fibers should also have not too rough surfaces and should be repositioned in specific directions to deserve the purpose of the new application. Finally the cost of the recovered fibers represents the main barrier for implementing these processes on an industrial scale, as pristine fibers remains less expensive. A simpler transformation of the composite material is to shred it. This solution represents however a significant down-cycling of the composite material, as the resulting shredded composite can only be used as a filler or similar material. Neowa GmbH, is using this technique to recycle glass fiber reinforced thermosetting composites in cement production. This company is currently the only industrial recycling station in Europe able to process composites. Another solution based on shredded composite was developed by Miljøskærm in Denmark. The company uses agglomerated shredded composite in sound insulation panels. Given the challenges presented, it seems that a sustainable solution for recycling wind turbine blade will need to be based on a combination of several solutions in order to consider all possible scenarios.

Working Principle of Wind Turbine

There is an air turbine of large blades attached on the top of a supporting tower of sufficient height. When wind strikes on the turbine blades, the turbine rotates due to the design and alignment of rotor blades. The shaft of the turbine is coupled with an electrical generator. The output of the generator is collected through electric power cables.

When the wind strikes the rotor blades, blades start rotating. The turbine rotor is connected to a high-speed gearbox. Gearbox transforms the rotor rotation from low speed to high speed. The high-speed shaft from the gearbox is coupled with the rotor of the generator and hence the electrical generator runs at a higher speed. An exciter is needed to give the required excitation to the magnetic coil of the generator field system so that it can generate the required electricity. The generated voltage at output terminals of the alternator is proportional to both the speed and field flux of the alternator. The speed is governed by wind power which is out of control. Hence to maintain uniformity of the output power from the alternator, excitation must be controlled according to the availability of natural wind power. The exciter current is controlled by a turbine controller which senses the wind speed. Then output voltage of electrical generator(alternator) is given to a rectifier where the alternator output gets rectified to DC. Then this rectified DC output is given to line converter unit to convert it into stabilized AC output which is ultimately fed to either electrical transmission network or transmission grid with the help of step up transformer. An extra units is used to give the power to internal auxiliaries of wind turbine (like motor, battery etc.), this is called Internal Supply Unit.

There are other two control mechanisms attached to a modern big wind turbine:

- Controlling the orientation of the turbine blade.
- Controlling the orientation of the turbine face.

The orientation of turbine blades is governed from the base hub of the blades. The blades are attached to the central hub with the help of a rotating arrangement through gears and small electric motor or hydraulic rotary system. The system can be electrically or mechanically controlled depending on its design. The blades are swiveled depending upon the speed of the wind. The technique is called pitch control. It provides the best possible orientation of the turbine blades along the direction of the wind to obtain optimized wind power.

The orientation of the nacelle or the entire body of the turbine can follow the direction of changing wind direction to maximize mechanical energy harvesting from the wind.

The direction of the wind along with its speed is sensed by an anemometer (automatic speed measuring devices) with wind vanes attached to the back top of the nacelle. The signal is fed back to an electronic microprocessor-based controlling system which governs the yaw motor which rotates the entire nacelle with gearing arrangement to face the air turbine along the direction of the wind.

An internal Block diagram of a wind turbine:

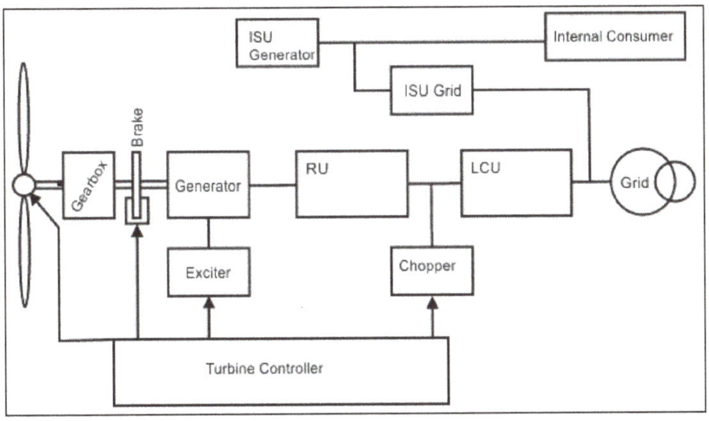

Wind Turbine Power Calculations

Wind turbines work by converting the kinetic energy in the wind first into rotational kinetic energy in the turbine and then electrical energy that can be supplied, via the national grid. The energy available for conversion mainly depends on the wind speed and the swept area of the turbine. When planning a wind farm it is important to know the expected power and energy output of each wind turbine to be able to calculate its economic viability.

Wind Turbine at North Wales coast.

Problem Statement

With the knowledge that it is of critical economic importance to know the power and therefore energy produced by different types of wind turbine in different conditions, in this exemplar we will calculate the rotational kinetic power produced in a wind turbine at its rated wind speed. This is the minimum wind speed at which a wind turbine produces its rated power.

Mathematical Model

The following table shows the definition of various variables used in this model:

- E = Kinetic Energy (J)
- m = Mass (kg)
- v = Wind Speed (m/s)
- P = Power (W)
- $\frac{dm}{dt}$ = Mass flow rate (kg/s)
- $\frac{dE}{dt}$ = Energy Flow Rate (J/s)
- ρ = Density (kg/m^3)
- A = Swept Area (m^2)
- C_p = Power Coefficient
- r = Radius (m)
- x = distance (m)
- t = time (s)

Under constant acceleration, the kinetic energy of an object having mass m and velocity v is equal to the work done W in displacing that object from rest to a distance s under a force F, i.e.:

$$E = W = Fs$$

According to Newton's Law, we have:

$$F = ma$$

Hence,

$$E = mas$$

Using the third equation of motion:

$$v^2 = u^2 + 2as$$

we get:

$$a = \frac{\left(v^2 - u^2\right)}{2s}$$

Since the initial velocity of the object is zero, i.e. u = 0, we get:

$$a = \frac{v^2}{2s}$$

Substituting it in equation E = mas, we get that the kinetic energy of a mass in motions is:

$$E = \frac{1}{2}mv^2$$

The power in the wind is given by the rate of change of energy:

$$P = \frac{dE}{dt} = \frac{1}{2}v^2\frac{dm}{dt}$$

As mass flow rate is given by:

$$\frac{dm}{dt} = \rho A\frac{dx}{dt}$$

and the rate of change of distance is given by:

$$\frac{dx}{dt} = v$$

we get:

$$\frac{dm}{dt} = \rho Av$$

Hence, from equation $P = \frac{dE}{dt} = \frac{1}{2}v^2\frac{dm}{dt}$, the power can be defined as:

$$P = \frac{1}{2}\rho Av^3$$

A German physicist Albert Betz concluded in 1919 that no wind turbine can convert more than 16/27 (59.3%) of the kinetic energy of the wind into mechanical energy turning a rotor. To this day, this is known as the Betz Limit or Betz' Law. The theoretical maximum power efficiency of any design of wind turbine is 0.59 (i.e. no more than 59% of the energy carried by the wind can be extracted by a wind turbine). This is called the "power coefficient" and is defined as:

$$C_{p\,\max} = 0.59$$

Also, wind turbines cannot operate at this maximum limit. The C_p value is unique to each turbine type and is a function of wind speed that the turbine is operating in. Once we incorporate various engineering requirements of a wind turbine - strength and durability in particular - the real world limit is well below the Betz Limit with values of 0.35-0.45 common even in the best designed wind turbines. By the time we take into account the other factors in a complete wind turbine system - e.g. the gearbox, bearings, generator and so on - only 10-30% of the power of the wind is ever actually converted into usable electricity. Hence, the power coefficient needs to be factored in equation, $P = \frac{1}{2}\rho A v^3$ and the extractable power from the wind is given by:

$$P_{avail} = \frac{1}{2}\rho A v^3 C_p$$

The swept area of the turbine can be calculated from the length of the turbine blades using the equation for the area of a circle:

$$A = \pi r^2$$

Where the radius is equal to the blade length as shown in the figure below:

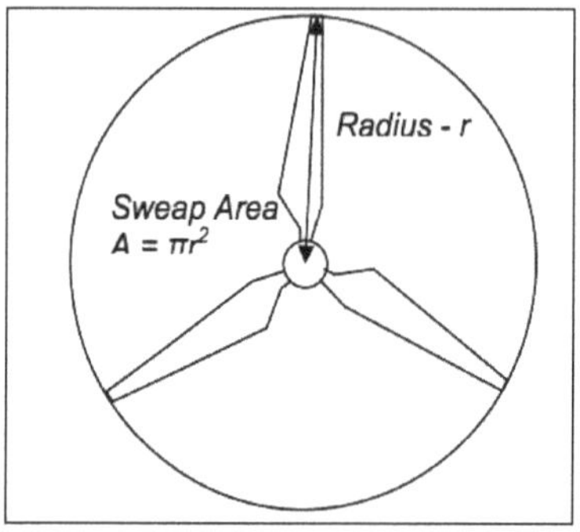

Small Wind Turbines Design

SWT design solutions include coupled numerical simulations (computational fluid dynamics (CFD), Finite Element Method (FEM), multi-body), investigating impact of the chosen design solutions on small wind turbine operation. It allowed objective evaluation of different design approaches (advantages and disadvantages), which in turn enabled the systematic identification of actual limitations as well as the opportunities for specific design solutions of SWTs: horizontal axis wind turbines (HAWTs) and vertical axis wind turbines (VAWTs); the rotor position in relation to the tower (upwind vs. downwind); and the addition of a duct that encapsulates the rotor (diffusor-augmented wind turbine—DAWT).

Horizontal and Vertical Axis SWTs

Wind turbines in general can be divided into two major groups based on the position of rotor axis of revolution: HAWTs and VAWTs.

HAWTs are currently the vast majority of installed solutions worldwide. Consequently, close to 75% of SWTs are HAWTs. That is due to their high efficiency in relation to low installation and maintenance costs. Apart from the unconventional solutions and one- and two-bladed configurations, the classic three-bladed layout is the default choice with a 99% market share of all operating HAWTs. Three blades is the lowest blade count which allows for easier management of centrifugal body forces. Some small turbines incorporate an increased amount of blades, which may improves aerodynamic efficiency slightly, but most often the added mass and material cost outweigh any benefits gained.

VAWTs are the earliest examples of harnessing the energy of the wind. These early machines were primarily based on a drag force as the mechanics to turn the turbine's working section. Some modern designs try to re-invent the concept, but the efficiency is generally unsatisfactory because of the limited tip speed ratio. Modern-day VAWTs most often use the aerodynamic principles of lift force to work, with Darrieus wind turbines being a good example. Some of the marketed advantages of VAWTs over the classic horizontal designs are as follows: easier maintenance thanks to the generator and the gearbox being placed near the ground, low cut-in wind speeds and no need for yaw mechanisms. In practice, maintenance and operational costs are similar because of the high loads on the bottom bearing, efficiency in converting wind kinetic energy into a mechanical power is relatively low for low wind speeds and response to an often changing wind direction dominated by turbulence is poor. Nevertheless, VAWTs are a popular choice for urban wind environments and are seeing some optimization research with advanced evolutionary algorithms aimed at raising the overall performance of these designs.

Shrouded SWTs

An interesting design solution for HAWT turbines is the addition of a circular duct that encapsulates the rotor. Such turbines have a number of commercial and scientific designations, such as DAWT, wind lens, compact wind acceleration turbine (CWAT).

Attempts to add a shroud in order to stabilize and accelerate the flow through a turbine have been reported nearly 150 years ago. An invention proposed by Ernest Bollée, patented in 1868, was an American style multiblade wind turbine with a stator in the form of a shroud. After decades of stagnancy, shrouded turbines have seen a major interest increase in recent years with many academic and industrial centers proposing mechanically and aerodynamically optimized solutions which are said to be vastly superior to non-shrouded designs. The influence of the shroud on the air flow has been proven to be beneficial in numerous wind tunnel tests, but little is available on the performance of such turbines in real outside conditions. What is more, the added diffusor is often expensive due to the amount of material used and the added mass puts more stress on the towers foundation and hinders the operation of the yaw system.

The concept of Framed Light Shell Diffuser was created as an attempt to lighten the diffuser made of glass fiber composites which are the most commonly used materials for such applications nowadays. Geometrical shape of the diffuser was developed based on CFD simulations considering turbine efficiency optimization for low wind speed conditions. The authors focused on developing an overall structural design, as well as detailed technical solutions (stiffer struts, connecting rods, material forming issues). Subsequently, static structural finite element analysis was performed in order to assess the stiffness and stress distribution based on load conditions in the form of pressure field form the CFD analysis and gravitational forces. Obtained results lead to a conclusion that designed aluminum shell frame diffuser may be an alternative for the composite diffuser. Moreover, the developed structure is characterized by lower mass and comparable stiffness to its composite counterpart. It is worth mentioning that lighter diffusers allow use of lighter supporting structures, such as towers or guy-wired masts. This fact may contribute to lowering the overall costs of future SWTs and could be very beneficial for the privately owned small wind turbine market.

Framed Light Shell Diffuser concept for SWT.

Upwind vs. Downwind SWTs

One of the fundamental distinguishing factors in horizontal wind turbines is the rotor position in relation to the tower. Today market is dominated by upwind rotors: relative to the wind, their rotor is located in front of the tower (windward), while downwind rotors are behind the tower (leeward).

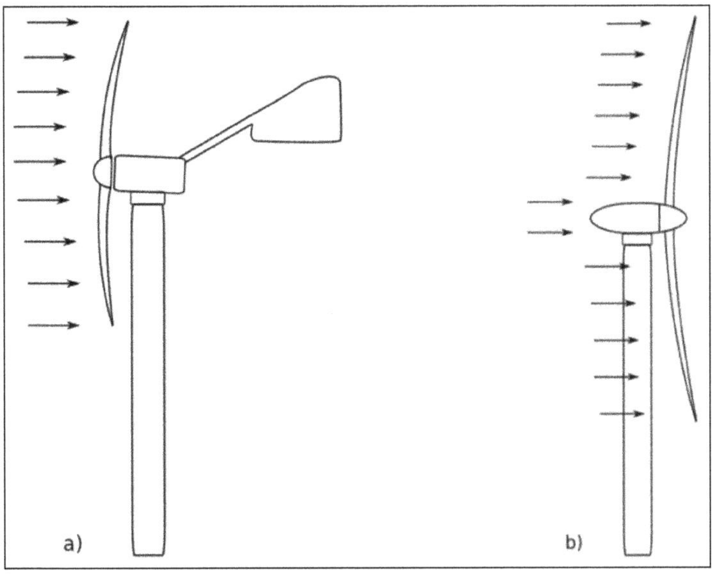

Upwind (a) and downwind (b) wind turbines.

In upwind design solution, the rotor is in the front of the unit (facing the wind) and it is characterized by higher efficiency due to the reduced tower impact on wind inducted at the working section. Unlike upwind, the downwind turbines experience a different inlet wind profile effectively changed by the development of a boundary layer on a nacelle and by the formation of wakes in the aerodynamic trail behind the mast. In upwind designs, the rotor needs to be made rather inflexible and placed at some distance from the tower to avoid collision. What is more, those are not self-aligning in the direction of the wind, and therefore, they need a tail vane or a yaw system to keep the rotor facing the wind.

In downwind turbines, the rotor is on the back side of the turbine (the lee side of the tower). Their main advantages are that they may be built without a yaw mechanism and the rotor may be made more flexible (since there is no danger of a tower strike). This may be an advantage both in regards to weight and the structural dynamics of the machine. Therefore, the basic advantage of the downwind machine is that theoretically, it may be built simpler and lighter than an upwind machine. Downwind turbines generally have lower aerodynamic efficiency, and the fluctuation in the wind power due to the rotor passing through the wind shade of the tower may give more fatigue loads on the turbine than those with an upwind design. Although the upwind type is more popular than the other one, the advantage that the downwind configuration can face the wind automatically makes them much more promising for SWTs due to its simplicity.

It is worth pointing out that the relative low cost of such research methods is a great advantage in comparison to real experiments (e.g., wind tunnel tests).

The first set of the obtained results concerns the dynamic response of the simulated variants in terms of rapidly changing wind speed conditions. Case one was here the IEC 61400–1 direction-change condition. Both considered turbine variants have proved to be capable of quick and precise nacelle turn during the changing direction, event and it can be stated that both design solutions work properly and similarly in this case. The upwind variant of the small wind turbine proves to have 5% higher efficiency than the downwind variant, which in turn resulted in a 13% higher energy output in economic analysis. For the 3 kW machine, the downwind variant should be at last 1200 USD cheaper than the upwind small wind turbine to be economically valid (by the application of lighter and cheaper materials, deflective blades, etc.); on the contrary, it is more likely to equip the small wind turbine in a tail vain having additional 1200 USD. The biggest disadvantage of the downwind design variant is the presence of significant fluctuations of momentums in rotor blades and forces on the top of turbine tower, which may cause a real danger of fatigue damage in the turbine construction as well as the risk of resonance.

Other SWTs Designs

Other design solutions that can be found in SWT applications or concepts are, that is, as follows: two-blade, single-blade turbines with a counterweigh, multiblade rotors, Venturi design turbines, Magnus effect turbines, multi-rotor solutions, H-type turbines, Darrieus–Savonius hybrid, Tornado/Jet/Vortex/Spiral airfoil, Fuller's design (bladeless—Tesla like), Pawlak's design, airborne concepts, and other unconventional solutions.

The main driver for multiblade (up to three) turbine development is the fact that aerodynamic efficiency increases with the number of blades. Increasing the number of blades from one to two yields a six percent increase in aerodynamic efficiency, whereas increasing the blade count from two to three yields only an additional three percent in efficiency.

Extraction of wind energy by a single rotor leaves a substantial amount of power unrecovered. To use this remaining potential, a two-stage wind turbine was proposed. Contra-rotating wind turbines possessing two co-axial rotors can theoretically gain up to 40% more energy from a given swept area as compared to a single-rotor turbine. Contra-rotating turbines require a generator tailored for this system in order to avoid expensive and impractical placement of the second generator on rotor-nacelle assembly (RNA) to convert energy from the additional rotor. In fact, the twin shaft technology of co-axial rotors presents a possibility to increase the rotation speed of the electrical generator by summing up the relative velocities of the rotor and stator. The main drawbacks of two-stage turbines are an increased interaction between the rotors posing problems from aero-mechanical point of view and the additional costs associated with the installation of the second stage and a more sophisticated generator.

The Savonius rotor is a self-starting, high-torque wind turbine. It may be used alone or to jump start the Darrieus rotor, a high-efficiency rotor, but with a limited capability to initiate operation on its own. This combination is presented as an effective design that combines the advantages of both designs.

The Magnus effect is the commonly observed effect in which a spinning ball (or cylinder) curves away from its principal flight path (a force perpendicular to the direction of movement, acting on the rotating cylinder or other rotary body, moving relative to the fluid). This makes a range of potential advantages with respect to traditional blade wind turbine. Radial cylinder location is analogous to wind wheel blades with horizontal axe. The basic advantages are said to be seen at low, but the most repeated wind velocities 2–6 m/s, at which blade wind turbines are not effective.

SWT's Blades

The blades are the components, which interact with wind and are designed to maximize the turbine efficiency. Blades are made from light materials, such as glass- or aluminum-based fiber-reinforced plastics, possessing good resistance to wear and tear. The fibers are incorporated in a matrix of polyester, epoxy resin, or vinyl ester constituting two shells kept together and strengthened by an internal matrix. The external surface of the blade is covered with a layer of colored gel to prevent ageing of composite material due to ultraviolet radiation.

A hollow shell corresponding to the defined blade envelope clearly provides a simple, efficient structure to resist flexural and torsional loads, and some blade manufacturers adopt this form of construction. The hollow shell structure defined by the airfoil section is not very efficient in resisting out-of-plane shear loads, so these are catered for by the inclusion of one or more shear webs oriented perpendicular to the blade chord.

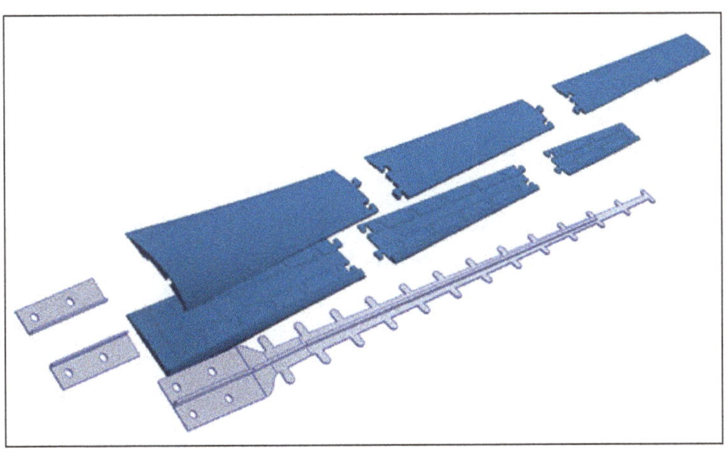

3D printed ABS polymer-based SWT blade with a stiffening core made of steel.

Five different materials were considered in terms of turbine blade optimization: steel, aluminum alloy, glass fiber composite, carbon fiber composite, and ABS polymer for three-dimensional (3D) printing. Blades made of all listed materials have fulfilled the strength criterion (stresses values generated by maximum loads did not crossed yield or fracture). Analysis revealed that the best material in terms of mass and stiffness is the carbon composite. Unfortunately, this material is very expensive and was opted with in order to keep SWT price reasonably low. Metal-based materials are characterized by very high stiffness, which reduces the blade tip deflection; however, the mass of the blade was too high in this case. What is more, a relatively high stiffness was a direct cause of relatively high stress values. Those values have raised justified questions on the fatigue toughness of the SWT. The ABS polymer is the cheapest and the lightest one of all mentioned. It is also the best choice from the manufacturing point of view. Unfortunately, it is also the most flexible material, producing high blade tip deflections even after the introduction of stiffening core made of steel.

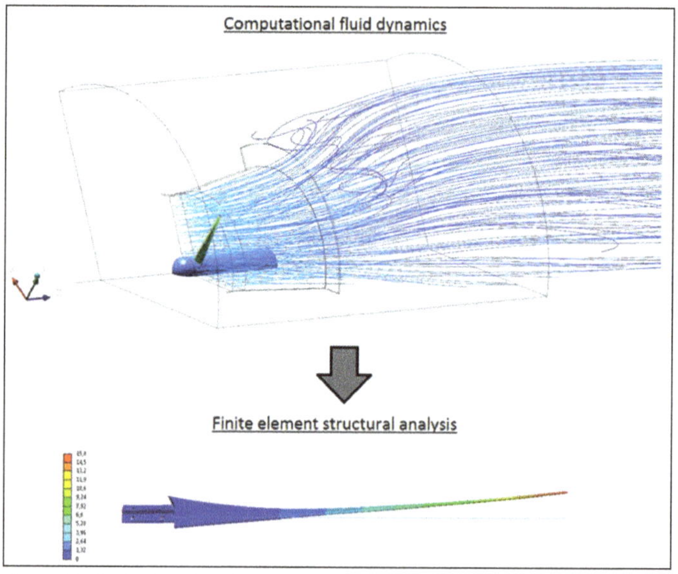

SWT blade deflection (ABS 3D printed coating with a steel core).

SWT's Generators

SWTs are usually equipped with synchronous generators operating at variable angular speeds based on permanent magnets (e.g., neodymium magnets). It is a design solution which can contribute to meeting the requirements of efficient operation. Working principles of the variable speed generator determine many other aspects of the wind turbine operation including, for example, a tremendously increased annual energy production (AEP) in comparison with equal-power fixed pitch, fixed speed turbines.

SWTs typically operate in autonomous systems without connection to the grid, in which they directly power heating and other loads that do not require stable electric parameters, dictated by economic issues. Potential expenditure related to connecting the SWT to the electrical grid exceeds the usual financial benefits of possible income from selling energy produced. The other aspect is the fact that SWTs are generally simple devices and they are rarely equipped with sophisticated control systems such as blade pitch mechanisms.

This type of generator produces alternating current (AC), which must be rectified to direct current (DC) by means of a simple bridge rectifier. Similar to solar photovoltaic (PV) systems, the DC voltage allows the use of these turbines for battery charging. For in-battery charging systems, a charge controller is added to prevent the battery from overcharging. A dump load is required to protect the inverter from overvoltage and to prevent the turbine from overspeeding.

SWT's Supporting Structures

The SWT is usually placed on a pole, preferably higher than 15 m, to lessen the impact of turbulent, sheared winds forming in close proximity to the ground. Utility-scale wind turbine towers are mostly of tubular type. In case of SWTs, tubular type towers are also popular; however, guyed lattice towers are also used. The tilt-up poles/towers are very popular since they are easy to install and offer good accessibility for maintenance and repair. Tubular towers are usually made of rolled steel, although sometimes reinforced concrete is used.

Roof-mounted SWT equipped with the diffuser.

Rooftop-mounted turbines are yet another option. Such turbines are directly installed on a building. Both vertical and horizontal axis turbines can be mounted in such a way, but they might be subject to slightly different wind profiles being inducted at the rotor. Particularly important for roof-mounted turbines is the identification of flow separations due to rooftop edges. Such flow structures not only influence the production efficiency but may be a source of unwanted loads that are difficult to control.

Classic Barrel Design of Savonius Wind Turbine

Its most basic design is S-shaped with two blades.

The Savonius turbine uses drag to push the curved blades to generate a torque that will make the rotor turn. Aerodynamically it is the simplest wind turbine to design and build which reduces its cost drastically compared to the aerofoil blade designs of the other VAWTs and HAWTs.

Its working principle is extremely simple. The turbine rotates because of the difference of the drag force acting on the concave and convex parts of its blades. Figure illustrates this principle:

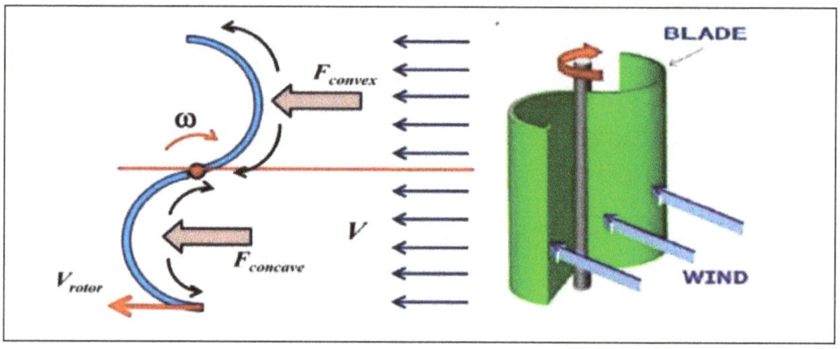

Working principle of a Savonius rotor.

The air is trapped in the concave part and pushes the turbine. The flow that hits the convex part does produce a drag that is lower than the one on the concave part. It is the differential of the drag force that causes this turbine to rotate.

This lowers the efficiency of the turbine as some of the wind's power is used in pushing the convex part and is hence "wasted". More blades can be added to the S shape design, and the same principle causes it to spin as shown in figure:

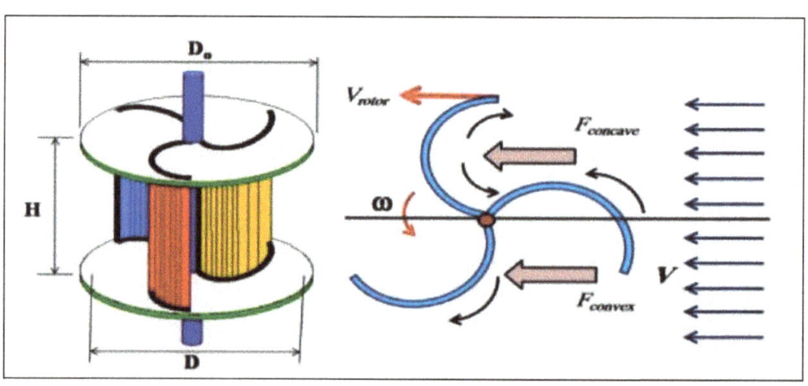

Three blades Savonius turbine.

According to, "a Savonius rotor requires 30 times more surface for the same power as a conventional rotor blade wind-turbine. Therefore it is only useful and economical for small power requirements." This makes Savonius ideal for small applications with low wind speeds. Savonius are hence desirable for their reliability, as they are able to work at several magnitudes of wind speed.

Characterization of Savonius Wind Turbines

Every Savonius wind turbine is characterized by the swept area A_s. This area influences the energy output of the turbine, and the larger it is, the more energy the turbine collects. $A_s = H * D$ where H is the height of the turbine and D is its diameter.

The tip speed ratio of the rotor is defined by the equation:

$$\lambda = \frac{V_{rotor}}{V} = \frac{\omega * d}{V}$$

where V is the wind speed, ω is the angular velocity of the turbine, and d is the diameter of the semicylindrical blade.

The torque coefficient C_t is the ratio between the torque in the rotor and the theoretical torque that the wind can cause: $C_t = \frac{T}{T_w} = \frac{T}{\frac{1}{4}\rho * A_s * d * V^2}$ where T is the torque in the rotor and ρ is the air density.

The static torque coefficient C_{ts} expresses the turbines ability to self-start. It is the ratio of the maximum static torque in the turbine and the theoretical wind torque:

$$C_{ts} = \frac{T_s}{T_w} = \frac{T}{\frac{1}{4}\rho * A_s * d * V^2}$$ Where T_s is the maximum static torque.

The torque in the rotor can be calculated using the following equation: $T = I * \alpha$, where I is the rotor's moment of inertia and α is the rotor's angular acceleration.

The power coefficient C_p is the ratio of the extracted power from the wind to the available power in the wind: $C_p = \frac{P_\omega}{P_a} = \frac{T * \omega}{\frac{1}{2}\rho * H * D * V^3}$

Using these factors, we can learn about the turbine's characteristics and analyze its performance.

Two Blades vs. Three Blades

Experimental studies by confirm that Savonius wind turbines do perform well at low wind speeds (cut in speed at around 2.5 m/s). According to the same study, two blades perform better than three blades as more drag is wasted in the three blades versions. The power coefficient of the two blade design is higher than that of the three blade design.

This result is confirmed by. In their experimental study, they found that 2 blades do perform better than 3 blades. They also found that 4 blades perform even better than 2 blades at low tip speed ratio (TSR), and that 3 blades perform better at higher TSR.

Wind Speed Classification

As we keep mentioning high vs. low wind speed, it is important to define what qualifies as high or low speeds.

The IEC 61400 is a comprehensive report that provides detailed comprehensive international standards for wind turbines by the International Electrotechnical Commission.

Table: Wind speed classification.

Wind Turbine Class		I	II	III
V_{ave} (M/S)		10	8.5	7.5
V_{ref} (M/S)		50	42.5	37.5
$V_{50,Gust}$ (M/S)		70	59.5	52.5
I_{ref}	A	0.16		
	B	0.14		
	C	0.12		

Where “V_{ave} is the annual mean wind speed at hub height. V_{ref} is the 50-year extreme wind speed over 10 minutes. V50, gust is the 50-year extreme gust over 3 seconds. I_{ref} is the mean turbulence intensity at 15 m/s, and A, B and C are the categories of higher, medium and lower turbulence intensity characteristics respectively”.

3D Computer Design

Savonius Wind Turbine: Classic Barrel Design

As the goal is to test the Icewind design, we will first design a classic barrel Savonius wind turbine, and then later modify it to reach the Icewind design. In the simulation phase, we will first obtain results from the tests on the classic barrel Savonius rotor, and then compare them to the new Icewind design.

The design of a barrel rotor is simple. The basic idea behind the simple two blades Savonius is that the blades are half cylinders (half a barrel, hence the name). These barrels do not often meet at the axis, but are set far apart as shown in figure, to create an overlap:

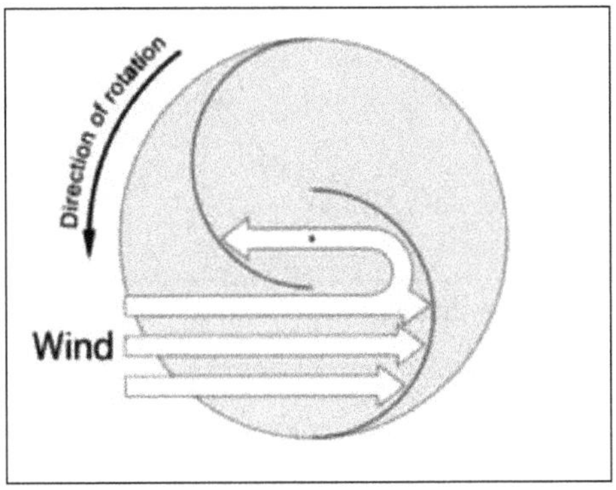

Overlap of the two blades on a Savonius rotor.

As the Icewind design does not have any overlap, we will not have this feature in our barrel Savonius design.

Many variables determine the shape of a Savonius barrel design. The ones we will be using are as follows - Do: outer diameter of the base of the rotor; D The distance between the two opposite blades; r the radius of the blades; and h the height of the rotor; we will be using a ratio D:h of 1:1. The ratio is the most common design feature of these turbines as they tend to be a square box shape, hence why the 1:1 ratio.

Another ratio we will be using is the Do:D, which determines how far the base extends beyond the blade. We will be using a Do:D of 1.1:1, which is the ratio used in the design by.

The following figures show the design we built using SolidWorks and the measurements used for two, three, and four blades barrel Savonius wind turbine:

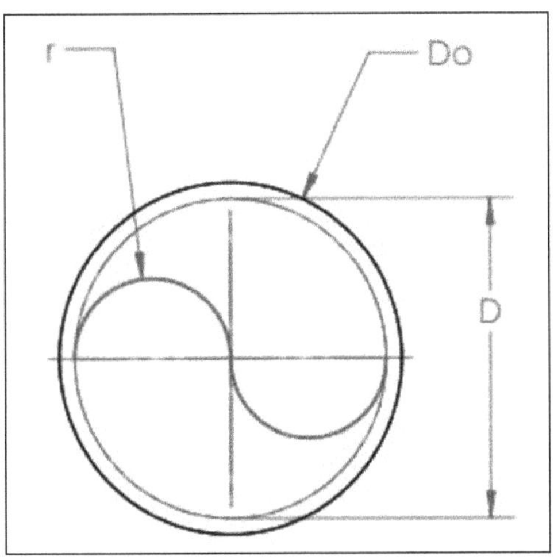

Two blades Savonius rotor design.

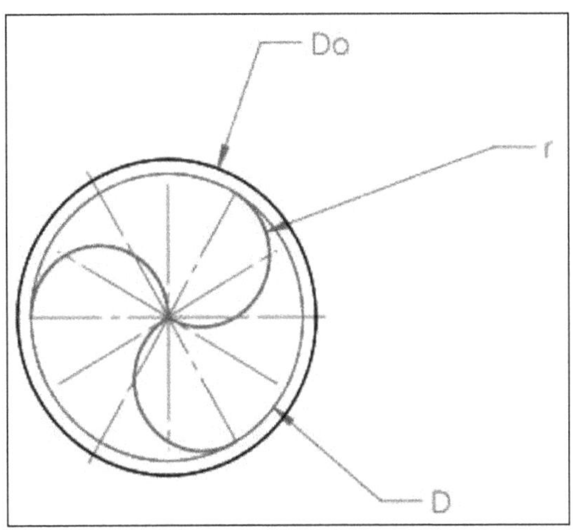

Three blades Savonius rotor design.

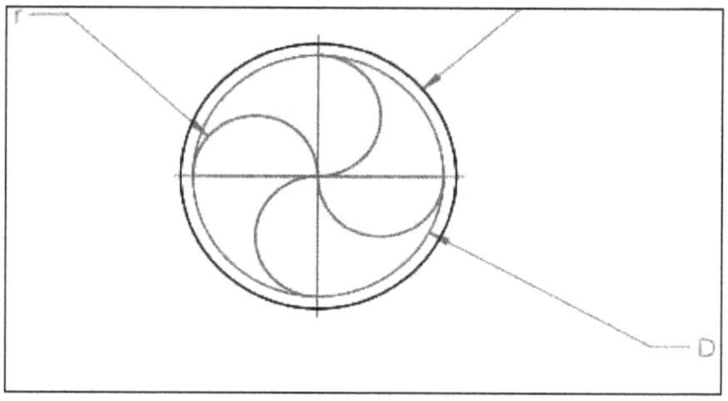

Four blades Savonius rotor design.

Table: Measurements (m).

Do	0.44
D	0.4
h	0.4
r	0.2

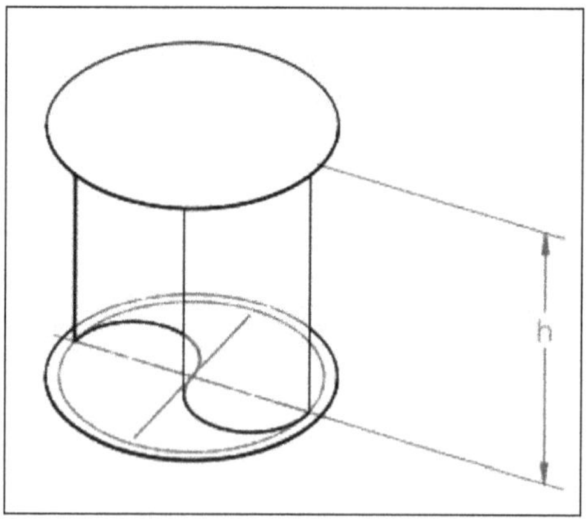

Two blades Savonius rotor design.

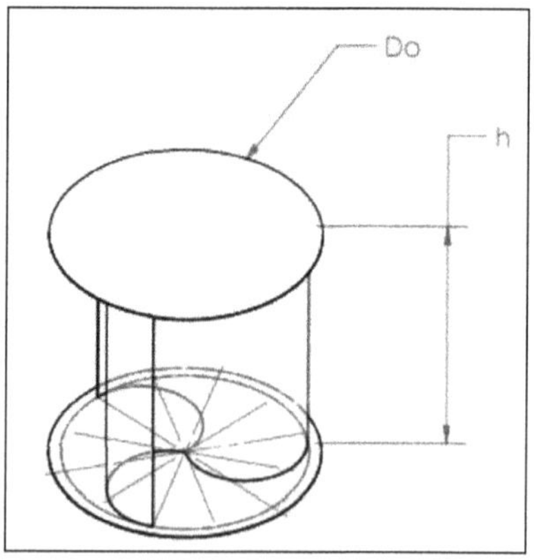

Three blades Savonius rotor design.

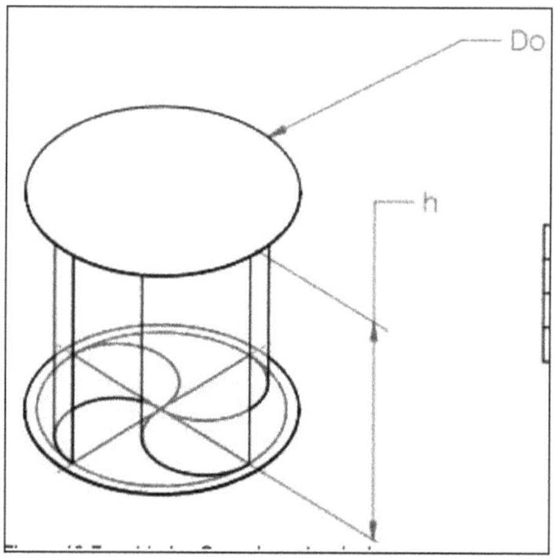

Four blades Savonius rotor design.

Table: Measurements (m).

Do	0.44
D	0.4
h	0.4
r	0.2

Two blades Savonius rotor design.

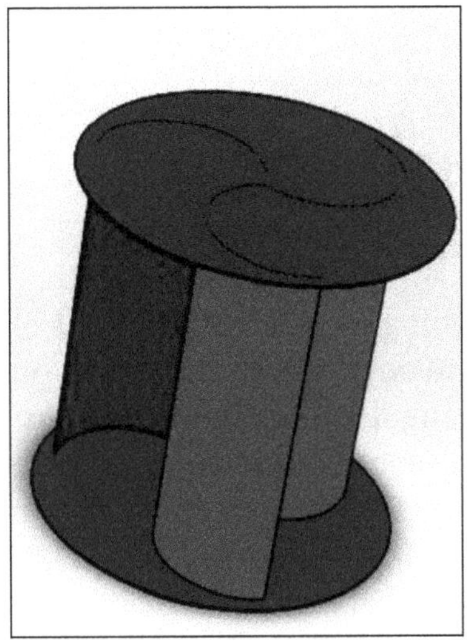

Three blades Savonius rotor design.

Four blades Savonius rotor design.

The base of the turbine is necessary for structural integrity. It provides support for the blades against the drag force of the wind pushing the blades of the rotor. As we will see in the coming section, the Icewind design does not use a base. In order for our comparison from the simulations to be representative, we will remove the base and connect the blades with a central axis. This will not affect our simulation since we are not performing a stress or a fatigue analysis.

The design is shown in the Figures below:

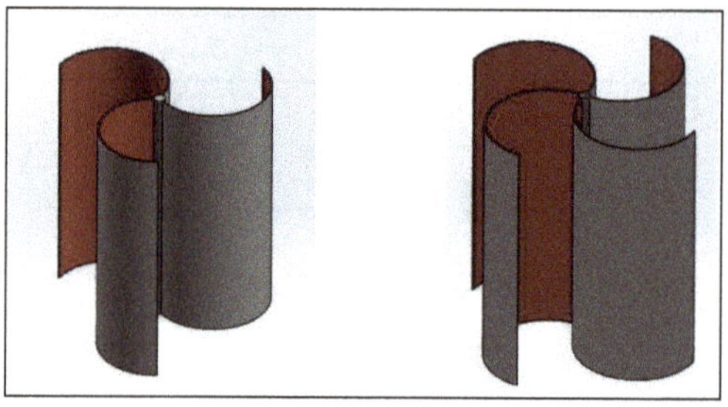

No base barrel design.

Icewind Design

Designing the Blade

The first thing we noticed is that the blades do not form a half circle at their base, but an incomplete one. This meant that in our design starting from the barrel shape we would cut a rectangular block from the inner side of the blade to make it similar to that of Icewind. The second distinctive feature is that the **D:h** ratio is not 1. We noticed that their blades are quite wide and that the diameter of the complete half circle (D/2) making the blade is the same as the height h (**D:h** close to 2). Once one end of the blade is trimmed to make the shape seen on their designs this ratio appears larger as the rotor blades are slightly less wide than the rotor is high. We achieved this ratio by trimming 1/10 of the blade on the inner end as will be demonstrated.

The outer curves on the blades are cutouts of tangent circles of unknown radius. Using SolisWorks constraint tools we managed to find the radius that will satisfy the position of the tip of the blade (1/5 of the blade's diameter) as well as tangencies to the lower parts of the blades.

This is shown in the following steps:

1. We start from a normal Savonius blade with 1:1 **D:h** ratio. We add the circles and solve for the mentioned constraints:

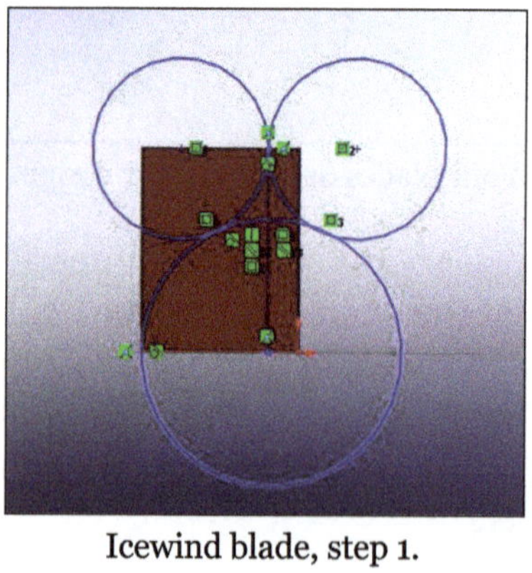

Icewind blade, step 1.

2. We trim the arcs that are needed and define shape of the blade:

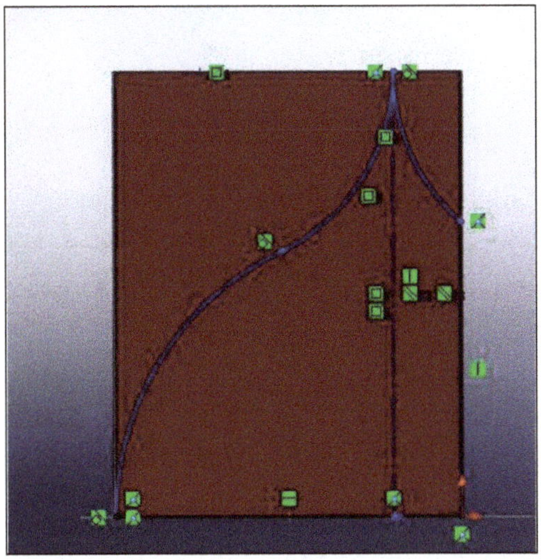

Icewind blade, step 2.

3. We intersect the created shape with original blade, and keep the intersection only:

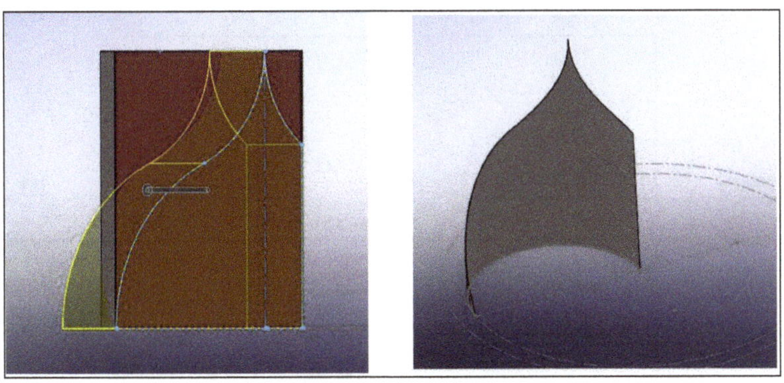

Icewind blade, step 3.

4. We trim 1/10 off the inner end of the blade:

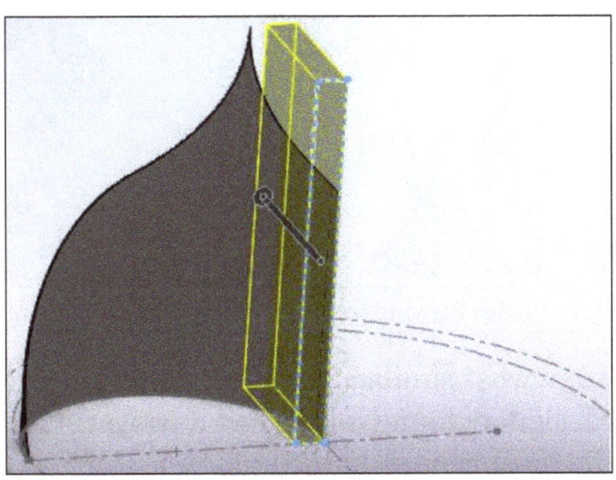

Icewind blade, step 4.

The final blade and its detailed measurements are shown in the figure below:

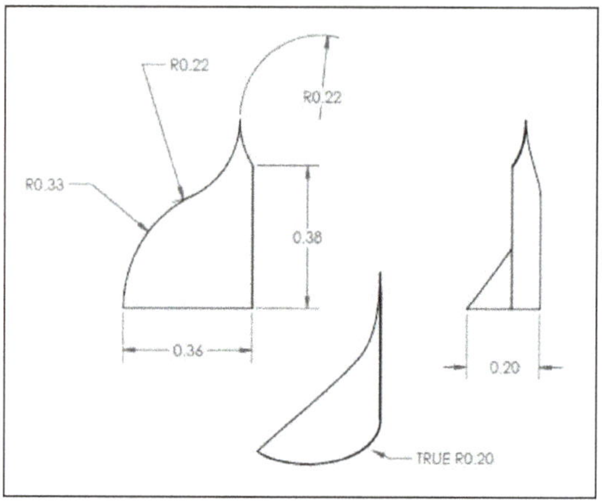

Icewind blade - Final design.

Now that we have the blade, designing the 3 and 4 blade rotors involved just attaching the blades to an axis.

The designs are shown below:

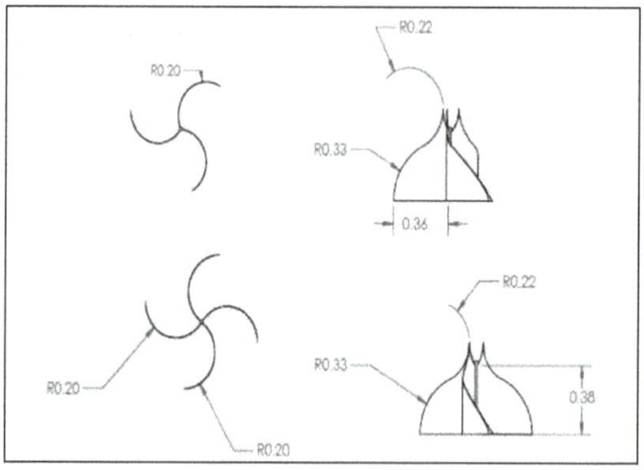

Icewind rotor design.

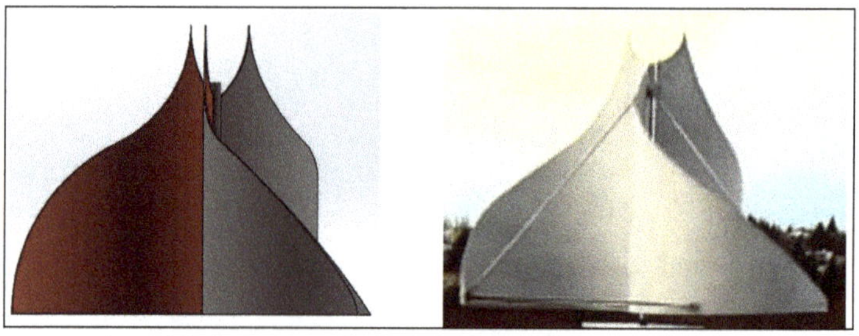

Three blades Icewind rotor: our design vs. the original.

The two designs look somewhat similar. The differences are due to the fact that every picture of the Icewind design we found online had its own characteristics. Consequently, we tried to match the most general and common features between all the different versions we found.

In order for our comparison to be accurate, we have to control for some variables (otherwise, the results of the analysis cannot be compared). Although we started from the barrel Savonius and modified it to make the Icewind design, the initial (barrel) and final design (Icewind) have no common ratios. To solve this problem we fixed the height of the two rotors, and modified the Icewind to have the same surface area as the barrel design. As the diameter of the Icewind design is double that of the barrel design, removing half of the blade compensated for the increase in diameter and kept the surface area constant at 0.12 m². By modifying, the area cut off the tip of the blade we increased the similarity in surface area to 99.86%, at a surface area of: 0.12441 m² for the barrel design, and 0.12459 m² for the Icewind design.

Theory of Wind Turbine and Betz Coefficient

For determining power extracted from wind by wind turbine we have to assume an air duct as shown in the figure. It is also assumed that the velocity of the wind at the inlet of the duct is V_1 and velocity of air at the outlet of the duct is V_2. Say, mass m of the air is passed through this imaginary duct per second.

Now due to this mass the kinetic energy of wind at the inlet of the duct is,

$$E_1 = \frac{1}{2} m V_1^2$$

Similarly, due to this mass the kinetic energy of wind at the outlet of the duct is,

$$E_2 = \frac{1}{2} m V_2^2$$

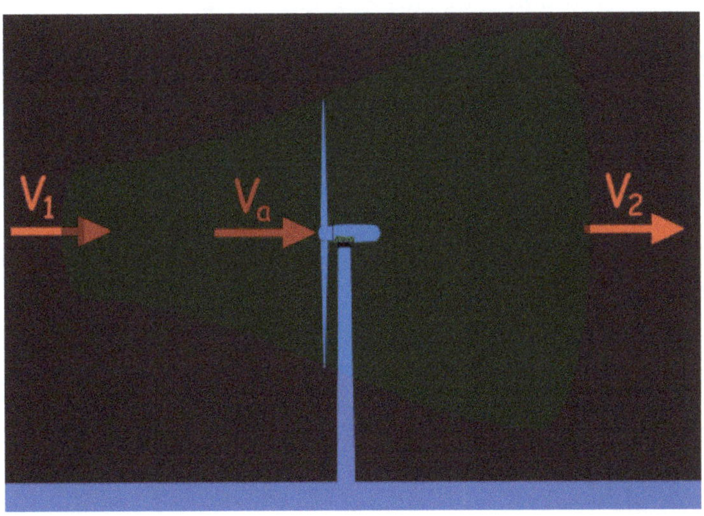

Hence, the kinetic energy of wind changed, during the flow of this quantity of air from the inlet to the outlet of the imaginary duct is,

$$E_1 - E_2 = \frac{1}{2} m V_1^2 - \frac{1}{2} m V_2^2$$

$$\Rightarrow E_1 - E_2 = \frac{1}{2} m \left(V_1^2 - V_2^2\right)$$

As we already said that, mass m of the air is passed through this imaginary duct in one second. Hence the power extracted from the wind is the same as the kinetic energy changed during the flow of mass m of the air from the inlet to the outlet of the duct.

We define power as the change of energy per second. Hence, this extracted power can be written as,

$$P_k = \frac{1}{2} m \left(V_1^2 - V_2^2\right)$$

As mass m of the air passes in one second, we refer the quantity m as the mass flow rate of the wind. If we think of that carefully, we can easily understand that mass flow rate will be the same at the inlet, at the outlet and as well as at every cross-section of the air duct. Since, whatever quantity of air is entering the duct, the same is coming out from the outlet.

If Va, A and ρ are the velocity of the air, the cross-sectional area of the duct and density of air at the turbine blades respectively, then the mass flow rate of the wind can be represented as,

$$m = \rho . A . V_a$$

$$mass = density \times volume = density \times length \times area$$

$$mass\ flow\ rate = \frac{\text{mass}}{\text{second}} = \frac{density \times length \times area}{\text{second}} = density \times \frac{\text{length}}{\text{second}} \times area$$

$$= density \times velocity \times area$$

Now, replacing m by $\rho V_a A$ in equation $P_k = \frac{1}{2} m \left(V_1^2 - V_2^2\right)$ we get,

$$P_k = \frac{1}{2} \rho A V_a \left(V_1^2 - V_2^2\right)$$

Now, as the turbine is assumed to be placed at the middle of the duct, the wind velocity at turbine blades can be considered as average velocity of inlet and outlet velocities.

$$Hence, V_a = \frac{V_1 + V_2}{2}$$

Now putting, this value of V_a *in equation* $P_k = \frac{1}{2}\rho A V_a\left(V_1^2 - V_2^2\right)$, *we get,*

$$P_k = \frac{1}{2}\rho A \frac{V_1 + V_2}{2}\left(V_1^2 - V_2^2\right)$$

$$= \frac{1}{2}\rho A\left(V_1^3 - V_2^3 - V_1 V_2^2 + V_1^2 V_2\right)$$

$$= \frac{1}{4}\rho A V_1^3\left(1 - \frac{V_2^3}{V_1^3} - \frac{V_2^2}{V_1^2} + \frac{V_2}{V_1}\right)$$

To obtain maximum power from wind, we have to differentiate equation $= \frac{1}{4}\rho A V_1^3\left(1 - \frac{V_2^3}{V_1^3} - \frac{V_2^2}{V_1^2} + \frac{V_2}{V_1}\right)$ in respect of V_2 and equate it to zero. That is,

$$\frac{dP_k}{dV_2} = \frac{1}{4}\rho A V_1^3\left(-3\frac{V_2^2}{V_1^3} - 2\frac{V_2}{V_1^2} + \frac{1}{V_1}\right) = 0$$

$$\Rightarrow \frac{1}{4}\rho A\left(-3V_2^2 - 2V_2 V_1 + V_1^2\right) = 0$$

$$\Rightarrow \frac{1}{4}\rho A\left(-3V_2^2 - 3V_2 V_1 + V_2 V_1 + V_1^2\right) = 0$$

$$\Rightarrow \frac{1}{4}\rho A\left[-3V_2\left(V_2 + V_1\right) + V_1\left(V_2 + V_1\right)\right] = 0$$

$$\Rightarrow \frac{1}{4}\rho A\left(-3V_2 + V_1\right)\left(V_2 + V_1\right) = 0$$

Here, ρ, A and ($V_2 + V_2$) cannot be zero, therefore,

$$\left(-3V_2 + V_1\right) = 0 \Rightarrow \frac{V_2}{V_1} = \frac{1}{3}$$

Now, Putting this value of V_2/V_1 in equation $\frac{1}{4}\rho A V_1^3\left(1 - \frac{V_2^3}{V_1^3} - \frac{V_2^2}{V_1^2} + \frac{V_2}{V_1}\right)$ we get,

$$P_k = \frac{1}{4}\rho A V_1^3\left(1 - \frac{1}{3^3} - \frac{1}{3^2} + \frac{1}{3}\right)$$

$$P_k = 0.5925 \times \frac{1}{2}\rho A V_1^3.$$

Betz Coefficient

From, the above equation it is found that the theoretical maximum power extracted from the wind is in the fraction of 0.5925 of its total kinetic power. This fraction is known as the Betz Coefficient. This calculated power is according to theory of wind turbine but actual mechanical power received by the generator is lesser than that and it is due to losses for friction rotor bearing and inefficiencies of aerodynamic design of the turbine.

From equation $P_k = 0.5925 \times \frac{1}{2} \rho A V_1^3$ it is clear that the extracted power is:

- Directly proportional to air density ρ. As air density increases, the power of the turbine increases.
- Directly proportional to the swept area of the turbine blades. If the length of the blade increases, the radius of the swept area increases accordingly, so turbine power increases.
- Turbine power also varies with velocity of the wind. That indicates if the velocity of wind doubles and the turbine power will increase to eight folds.

Aerodynamic Design of Horizontal-axis Wind Turbine Blades

Energy prices, supply uncertainties, and environmental concerns are driving the developed nations to rethink their energy mix and develop diverse sources of clean, renewable energy. Wind power, as an alternative to fossil fuels, is growing at the rate of 19.2% annually, with a worldwide installed capacity increased from 432,680MW in 2015 to 486,790MW at the end of 2016. All wind turbines installed by the end of 2016 worldwide can generate around 5% of the world's electricity demands. Wind power is widely used in Europe, Asia, and the United States. China, United States, Germany, India and Spain respectively are top five countries in terms of installed cumulative capacities at the end of 2016.

Wind power is also capable of becoming a major contributor to the developing nations of Africa. According to 2016 Global Wind Energy Council (GWEC) statistics, all wind turbines installed in Africa had a capacity of 3,906MW. Currently, South Africa is leading the continent with installed capacity of 1,471MW followed by Egypt with 810MW, Morocco with 787MW and Ethiopia with 324MW. Other promising countries in Africa include Tunisia, Kenya, Tanzania, where wind project development is firmly underway.

Wind energy application in Ethiopia has been limited to water pumping in the past. There is now, however, definite plan to exploit wind for power production. According to the Ethiopian Electric Power Corporation (EEPCo), Ethiopia has a capacity of generating more than 10,000MW from wind. Considering this substantial wind resource in the country, the government has committed itself to generate power from wind plants by constructing eight wind farms with total capacities of 1,116MW. So far, two wind farms such as Ashegoda, Adama I and Adama II wind farms started producing 324MW. Some of the erected wind turbines in Adama I wind farm are shown in figure below:

Some of the erected wind turbines in Adama I wind farm in Ethiopia.

Wind turbine power production depends on the interaction between the rotor and the wind. The blades of wind turbine receive kinetic energy of the wind, which is then transformed to mechanical energy, electrical forms, depending on our end uses. The objective of this study is to develop a user-interface computer program for Horizontal-Axis Wind turbine (HAWT) blade design and power performance prediction using the Blade Element Momentum (BEM) theory.

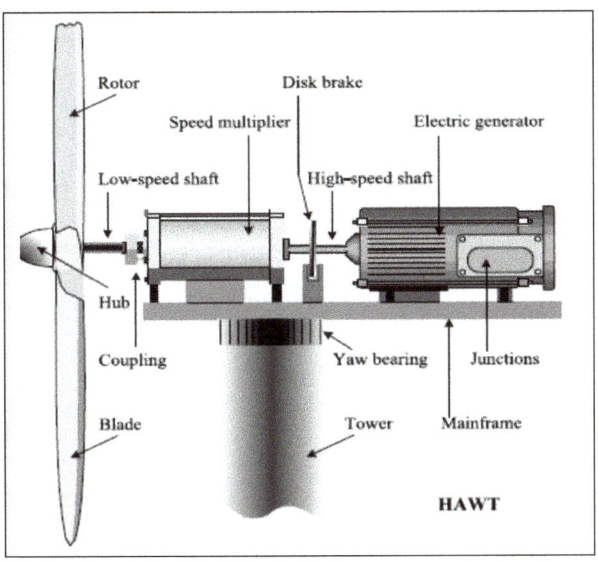

Main elements of HAWTs.

The primary objective in wind turbine design is to maximize the aerodynamic efficiency, or power extracted from the wind. However, the mechanical strength criteria, and economical aspects are also equally important and therefore must be considered during the design of wind turbines.The design of HAWT blades, to achieve satisfactory level of performance; starts with knowledge of the aerodynamic forces acting on the blades. In this topic, HAWT blade design from the aspect of aerodynamics is investigated. The structural design of HAWT blades is also as important as their aerodynamic design. The dynamic structural loads which a rotor will experience play the major role in determining the lifetime of the rotor. Obviously, aerodynamic loads are a major source of dynamic structural behavior, and ought to be accurately determined. In addition, the blade geometry parameters are required for dynamic load analysis of wind turbine rotors.

Aerodynamics of Hawts

Many factors play a role in the design of a wind turbine rotor, including aerodynamics, generator characteristics, blade strength and rigidity, noise levels, etc. But since wind energy conversion system is largely dependent on maximizing its energy extraction, rotor aerodynamics play important role in the minimization of the cost of energy. The rotor consists of the hub and blades of the wind turbine. These are often considered to be its most important components from both performance and overall cost standpoint.

The Actuator Disc Theory and the Betz Limit

A simple model, generally attributed to Betz can be used to determine the power from an ideal turbine rotor, the thrust of the wind on the ideal rotor and the effect of the rotor operation on the local wind field. The simplest aerodynamic model of a wind turbine is known as 'actuator disk model' in which the rotor is assumed like a homogenous disk that removes energy from the wind.

The theory of the ideal actuator disk is based on the following assumptions:

- Homogenous, incompressible, steady state fluid flow, no frictional drag;
- The pressure increment or thrust per unit area is constant over the disk;
- The rotational component of the velocity in the slipstream is zero;
- There is continuity of velocity through the disk;
- An infinite number of blades.

The analysis of the actuator disk theory assumes a control volume as shown in figure below.

In the control volume below, the only flow is across the ends of the stream tube. The turbine is represented by a uniform actuator disk which creates a discontinuity of pressure in the stream tube of air flowing through it. Note also that this analysis is not limited to any particular type of wind turbine.

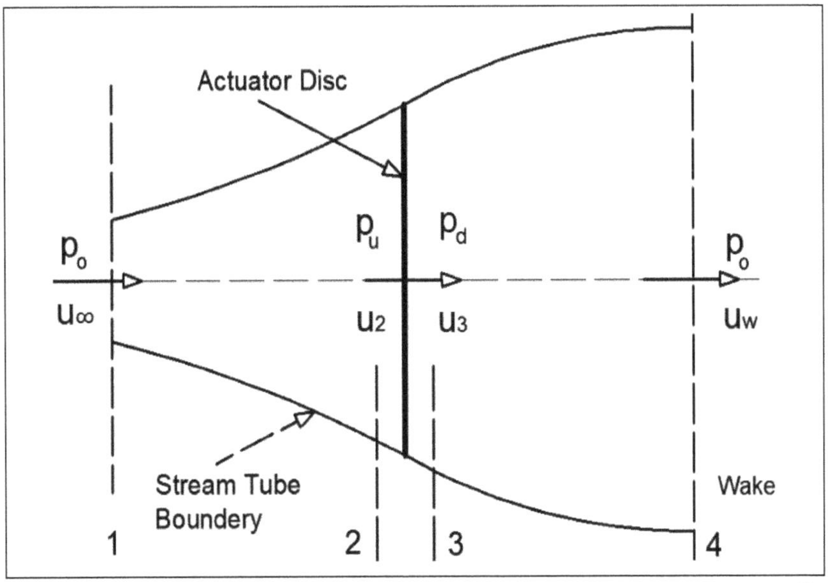

Idealized flow through a wind turbine represented by a non-rotating actuator disc.

From t elocity through the disk exists, the velocities are equal to the velocity at the rotor:

$$U_2 = u_3 = u_R$$

For steady state flow, air mass flow rate through the disk can be written as:

$$\dot{m} = \rho A u_R$$

the control volume enclosing the whole system, the net force can be found on the contents of the control volume. That force is equal and opposite to the thrust, T which is the force of the wind on the wind turbine. Hence from the conservation of linear momentum for a one-dimensional, incompressible, time-invariant flow the thrust is equal and opposite to the change in momentum of air stream:

$$T = -\dot{m}\left(u_\infty - u_w\right)$$

No work is done on either side of the turbine rotor. Thus the Bernoulli function can be used in the two control volumes on either side of the actuator disk. Between the free-stream and upwind side of the rotor and between the downwind side of the rotor and far wake respectively:

$$p_o + \frac{1}{2}\rho u_\infty^2 = p_u + \frac{1}{2}\rho u_R^{\,2}$$

$$p_d + \frac{1}{2}\rho u_R^2 = p_o + \frac{1}{2}\rho u_w^{\,2}$$

The thrust can also be expressed as the net sum forces on each side of the actuator disk:

$$T = Ap'$$

where,

$$p' = (p_u - p_d)$$

By using equations $p_o + \frac{1}{2}\rho u_\infty^2 = p_u + \frac{1}{2}\rho u_R^2$ and $p_d + \frac{1}{2}\rho u_R^2 = p_o + \frac{1}{2}\rho u_w^2$, the pressure decrease, p′ can be found as:

$$p' = \frac{1}{2}\rho\left(u_\infty^2 - u_w^2\right)$$

And by substituting equation $p' = \frac{1}{2}\rho\left(u_\infty^2 - u_w^2\right)$ in to equation T = Ap′,

$$T = \frac{1}{2}\rho A\left(u_\infty^2 - u_w^2\right)$$

By equating the thrust values from equation $T = -\dot{m}\left(u_\infty - u_w\right)$ in which substituting Equation $\dot{m} = \rho A u_R$ in place of $\dot{m}$ and equation $T = \frac{1}{2}\rho A\left(u_\infty^2 - u_w^2\right)$, the velocity at the rotor plane can be found as:

$$u_R = \frac{u_\infty + u_w}{2}$$

If an axial induction factor, *a* is defined as the fractional decrease in the wind velocity between the free stream and the rotor plane then,

$$a = \frac{u_\infty - u_R}{u_\infty}$$

$$u_R = u_\infty\left(1 - a\right)$$

Using equation $u_R = \frac{u_\infty + u_w}{2}$ and $u_R = u_\infty\left(1 - a\right)$ one can get:

$$u_w = u_\infty\left(1 - 2a\right)$$

The velocity and pressure distribution are illustrated in figure. Because of continuity, the diameter of flow field must increase as its velocity decreases and note that there occurs sudden pressure drop at rotor plane which contributes the torque rotating turbine blades.

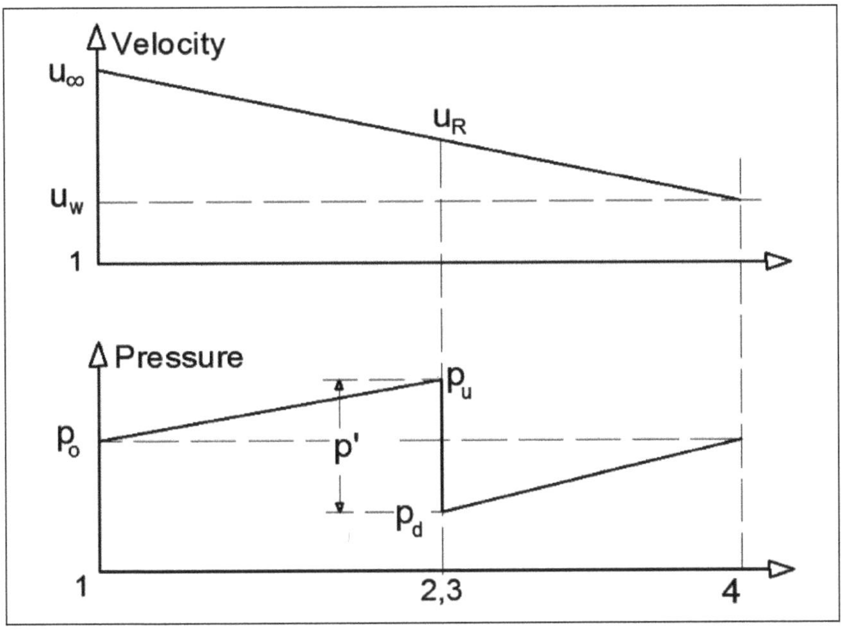

Velocity and Pressure distribution along the stream tube.

The power output, P is equal to the thrust times the velocity at the rotor plane:

$$P = Tu_R$$

Using equation $T = \frac{1}{2}\rho A\left(u_\infty^{\ 2} - u_w^{\ 2}\right)$,

$$P = \frac{1}{2}\rho A\left(u_\infty^{\ 2} - u_w^{\ 2}\right)u_R$$

And substituting for u_R and u_W from equations $u_R = u_\infty(1-a)$ and $u_w = u_\infty(1-2a)$ to equations $P = \frac{1}{2}\rho A\left(u_\infty^{\ 2} - u_w^{\ 2}\right)u_R$:

$$P = 2\rho Aa(1-a)^2 u_\infty^{\ 3}$$

The power performance parameters of a wind turbine can be expressed in dimensionless form, in which the power coefficient, C_P is given in the following equation:

$$C_P = \frac{P}{0.5\,\rho u_\infty^{\ 3} A}$$

Using equation $P = 2\rho Aa(1-a)^2 u_\infty^{\ 3}$ and $C_P = \frac{P}{0.5\,\rho u_\infty^{\ 3} A}$, the power coefficient C_P becomes:

$$C_P = 4a(1-a)^2$$

The maximum C_P is determined by taking the derivative of equation $C_P = 4a(1-a)^2$ with respect to a and setting it equal to zero yields:

$$\left(C_P\right)_{\max} = \frac{16}{27} = 0.5926$$

when, $a = 1/3$.

This result indicates that if an ideal rotor were designed and operated so that the wind speed at the rotor were 2/3 of the free stream wind speed, then it would be operating at the point of maximum power production. This is known as the Betz limit.

From equations $T = \frac{1}{2}\rho A\left(u_\infty^2 - u_w^2\right)$ and $u_w = u_\infty(1-2a)$ the axial thrust on the disk can be written in the following form:

$$T = 2\rho A a(1-a)u_\infty^2$$

Similar to the power coefficient, thrust coefficient can be defined by the ratio of trust force to dynamic force as shown in the following equation:

$$C_T = \frac{T}{0.5\rho u_\infty^2 A}$$

Using equation $T = 2\rho A a(1-a)u_\infty^2$ and $C_T = \frac{T}{0.5\rho u_\infty^2 A}$, the trust coefficient C_T becomes:

$$C_T = 4a(1-a)$$

Note that C_T has a maximum of 1.0 when a = 0.5 and the downstream velocity (wake velocity) is zero. At maximum power output ($a = 1/3$), CT has a value of 8/9. A graph of the power and trust coefficients for an ideal Betz turbine and the non-dimensionalized downstream wind speed are illustrated in the figure below.

As mentioned above, this idealized model is not valid for axial induction factors greater than 0.5. In practice, as the axial induction factor approaches and exceeds 0.5, complicated flow patterns that are not represented in this simple model result in thrust coefficients that can go as high as 2.0.

In conclusion, the actuator disk theory provides a rational basis for illustrating that the flow velocity at the rotor is different from the free-stream velocity. The Betz limit, $C_{P,max}$ =16/27, shows the maximum theoretically possible rotor power coefficient that can be attained from a wind turbine. In practice, three effects lead to a decrease in the maximum achievable power coefficient:

- Rotation of the wake behind the rotor;

- Finite number of blades and associated tip losses;
- Non-zero aerodynamic drag.

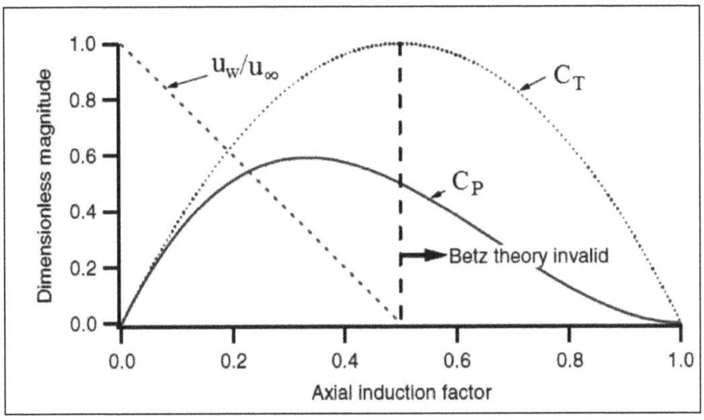

Operating parameters for a Betz turbine.

Note that the overall turbine efficiency is a function of both the rotor power coefficient and the mechanical (including electrical) efficiency of the wind turbine. Therefore:

$$P_{out} = \frac{1}{2}\rho A u_\infty^2 \left(\eta_{mech} C_P\right).$$

General Momentum Theory

In the previous analysis using linear momentum theory, it was assumed that no rotation was imparted to the flow. The previous analysis can be extended to the case where the rotating rotor generates angular momentum, which can be related to rotor torque. In the case of a rotating wind turbine rotor, the flow behind the rotor rotates in the opposite direction to the rotor, in reaction to the torque exerted by the flow on the rotor. An annular stream tube model of this flow, illustrating the rotation of the wake, is shown in figure below.

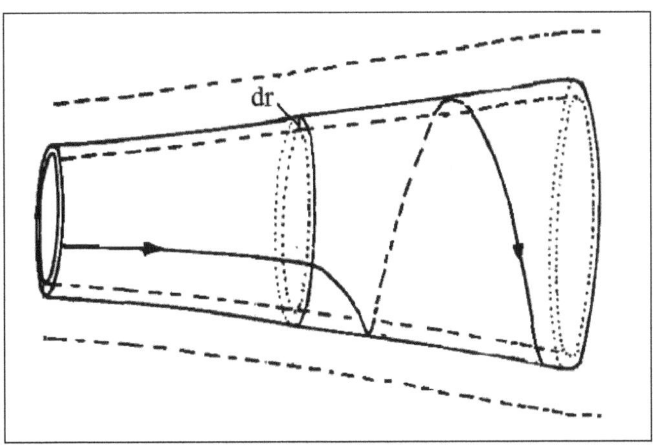

Stream tube model of flow behind rotating wind turbine blade.

The generation of rotational kinetic energy in the wake results in less energy extraction by the rotor than would be expected without wake rotation. In general, the extra kinetic energy in the wind turbine wake will be higher if the generated torque is higher. Thus, as will be shown here, slow-running wind turbines (with a low rotational speed and a high torque) experience more wake rotation losses than high-speed wind machines with low torque.

Figure gives a schematic of the parameters involved in this analysis. Subscripts denote values at the crosssections identified by numbers. If it is assumed that the angular velocity imparted to the flow stream, ω, is small compared to the angular velocity, Ω, of the wind turbine rotor, then it can also be assumed that the pressure in the far wake is equal to the pressure in the free stream . The analysis that follows is based on the use of an annular stream tube with a radius r and a thickness dr, resulting in a cross-sectional area equal to 2πrdr. The pressure, wake rotation, and induction factors are all assumed to be functions of radius.

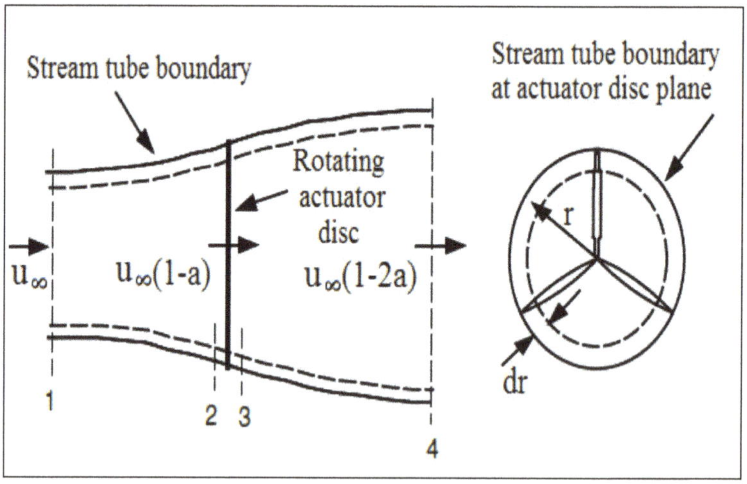

Geometry for rotor analysis.

If one uses a control volume that moves with the angular velocity of the blades, the energy equation can be applied in the sections before and after the blades to derive an expression for the pressure difference across the blades. Note that across the flow disc, the angular velocity of the air relative to the blade increases from Ω to Ω+ ω, while the axial component of the velocity remains constant. The results are:

$$p_2 - p_3 = \rho\left(\Omega + \frac{1}{2}\omega\right)\omega r^2$$

The resulting thrust on an annular element, dT, is:

$$dT = (p_2 - p_3)dA = \left[\rho\left(\Omega + \frac{1}{2}\omega\right)\omega r^2\right]2\pi r dr$$

An angular induction factor, a', is then defined as:

$$a' = \omega/2\Omega$$

Note that when wake rotation is included in the analysis, the induced velocity at the rotor consists of not only the axial component, $u_\infty a$, but also a component in the rotor plane, $r\Omega a'$.

The expression for the thrust becomes:

$$dT = 4a'(1+a')\frac{1}{2}\rho\,\Omega^2 r^2 2\pi r dr$$

Following the previous linear momentum analysis, the thrust on an annular cross-section can also be determined by the following expression that uses the axial induction factor, a, (note that u_1, the free stream velocity, is designated by u_∞ in this analysis):

$$dT = 4a(1-a)\frac{1}{2}\rho u_\infty^2 2\pi r dr$$

Equating the two expressions for thrust gives:

$$\frac{a(1-a)}{a'(1+a')} = \frac{\Omega^2 r^2}{u_\infty^{\;2}} = \lambda_r^2$$

where λ_r is the local tip speed ratio. This result will be used later in the analysis.

The tip speed ratio, λ, defined as the ratio of the blade tip speed to the free stream wind speed, is given by:

$$\lambda = \Omega R / u_\infty$$

The tip speed ratio often occurs in the aerodynamic equations for the rotor. The local speed ratio is the ratio of the rotor speed at some intermediate radius to the wind speed:

$$\lambda_r = \Omega r / u_\infty = \lambda r / R$$

Next, one can derive an expression for the torque on the rotor by applying the conservation of angular momentum. For this situation, the torque exerted on the rotor, Q, must equal the change in angular momentum of the wake. On an incremental annular area element this gives:

$$dQ = d\dot{m}(\omega r)(r) = (\rho u_2 2\pi r dr)(\omega r)(r)$$

Since $u_2 = u_\infty (1 - a)$ and $a' = \omega/2\Omega$, this expression reduces to:

$$dQ = 4a'(1-a)\frac{1}{2}u_\infty \Omega r^2 2\pi r dr$$

The power generated at each element, dP, is given by:

$$dP = \Omega dQ$$

Substituting for dQ in this expression and using the definition of the local speed ratio, λ_r, (equation $\lambda_r = \Omega r / u_\infty = \lambda r / R$), the expression for the power generated at each element becomes:

$$dP = \frac{1}{2}\rho A u_\infty^3 \left[\frac{8}{\lambda^2} a'(1-a)\lambda_r^3 d\lambda_r\right]$$

It can be seen that the power from any annular ring is a function of the axial and angular induction factors and the tip speed ratio. The axial and angular induction factors determine the magnitude and direction of the air flow at the rotor plane. The local speed ratio is a function of the tip speed ratio and radius.

The incremental contribution to the power coefficient, dC_p, from each annular ring is given by:

$$dC_P = \frac{dP}{1/2\rho A u_\infty^3}$$

Thus,

$$C_P = \frac{8}{\lambda^2}\int_0^\lambda a'(1-a)\lambda_r^3 d\lambda_r$$

In order to integrate this expression, one needs to relate the variables a, a', and λ_r. Solving equation $\frac{a(1-a)}{a'(1+a')} = \frac{\Omega^2 r^2}{u_\infty^2} = \lambda_r^2$ to express a' in terms of a, one gets:

$$a' = -\frac{1}{2} + \frac{1}{2}\sqrt{\left[1 + \frac{4}{\lambda_r^2}a(1-a)\right]}$$

The aerodynamic conditions for the maximum possible power production occur when the term $a'(1-a)$ in equation $C_P = \frac{8}{\lambda^2}\int_0^\lambda a'(1-a)\lambda_r^3 d\lambda_r$ is at its greatest value.

Substituting the value for *a'* from equation $a' = -\frac{1}{2} + \frac{1}{2}\sqrt{\left[1 + \frac{4}{\lambda_r^2} a(1-a)\right]}$ into *a'*(1-*a*)

and setting the derivative with respect to *a* equal to zero yields:

$$\lambda_r^2 = \frac{(1-a)(4a-1)^2}{1-3a}$$

This equation defines the axial induction factor for maximum power as a function of the local tip speed ratio in each annular ring. Substituting into equation $\frac{a(1-a)}{a'(1+a')} = \frac{\Omega^2 r^2}{u_\infty^2} = \lambda_r^2$, one finds that, for maximum power in each annular ring:

$$\frac{1 \quad 3}{4 \quad 1}$$

If equation $\lambda_r^2 = \frac{(1-a)(4a-1)^2}{1-3a}$ is differentiated with respect to *a*, one obtains a relationship between $d\lambda_r$ and da at those conditions that result in maximum power production:

$$2\lambda_r d\lambda_r = \left[6(4a-1)(1-2a)^2 / (1-3a)^2\right]$$

Now, substituting the equations $\lambda_r^2 = \frac{(1-a)(4a-1)^2}{1-3a}$ and $a' = \frac{1-3a}{4a-1}$ into the expression for the power coefficient (equation $C_P = \frac{8}{\lambda^2}\int_0^{\lambda} a'(1-a)\lambda_r^3 d\lambda_r$) gives:

$$C_{P.\max} = \frac{24}{\lambda^2}\int_{a_1}^{a_2}\left[\frac{(1-a)(1-2a)(1-4a)}{(1-3a)}\right]^2 da$$

Here the lower limit of integration, a_1, corresponds to the axial induction factor for $\lambda_r = \lambda_h = 0$ and the upper limit, a_2, corresponds to the axial induction factor at $\lambda_r = \lambda$. Also, from equation $\lambda_r^2 = \frac{(1-a)(4a-1)^2}{1-3a}$:

$$\lambda^2 = (1-a_2)(1-4a_2)/(1-3a_2)$$

Note that from equation $\lambda_r^2 = \frac{(1-a)(4a-1)^2}{1-3a}$, a_1 =0.25 gives λ_r a value of zero.

Equation $\lambda^2 = (1-a_2)(1-4a_2)/(1-3a_2)$ can be solved for the values of a_2 that correspond to operation at tip speed ratios of interest. Note also from equation $\lambda^2 = (1-a_2)(1-4a_2)/(1-3a_2)$, $a_2 = 1/3$ is the upper limit of the axial induction factor, *a*, giving an infinitely large tip speed ratio.

The definite integral in equation $C_{P.\max} = \frac{24}{\lambda^2}\int_{a_1}^{a_2}\left[\frac{(1-a)(1-2a)(1-4a)}{(1-3a)}\right]^2 da$ can be evaluated by changing variables: substituting x for (1−3*a*) in equation

$$C_{P.\max} = \frac{24}{\lambda^2}\int_{a_1}^{a_2}\left[\frac{(1-a)(1-2a)(1-4a)}{(1-3a)}\right]^2 da$$

and evaluating the integral, the result is:

$$C_{p.\max} = \frac{8}{729\lambda^2}\left\{\begin{matrix}\frac{64}{5}x^5 + 72x^4 + 124x^3 \\ +38x^2 - 63x \\ -12\ In\,x - 4x^{-1}\end{matrix}\right\}_{x=(1-3a_2)}^{x=0.25}$$

The result of general momentum theory are graphically represented in figure. which also shows the Betz limit of an ideal turbine based on the linear momentum analysis. The result shows that the higher the tip-speed ratio, the greater the maximum power coefficient. Note that figure, represents the numerical values for $C_{P,\max}$ as a function of λ, with corresponding values for the axial induction factor at the tip, a_2.

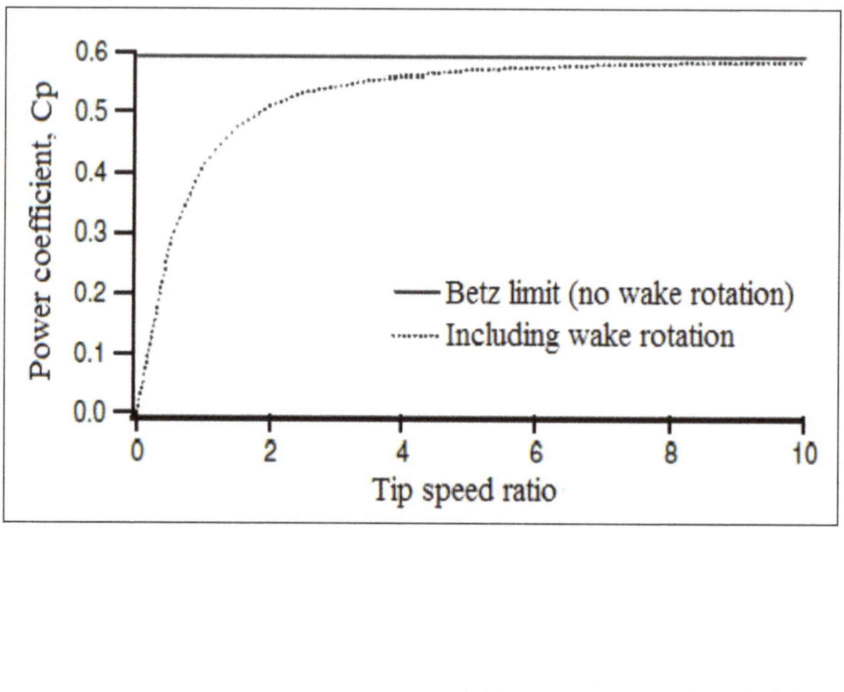

Theoretical maximum power coefficient as a function of tip speed ratio for an ideal HAWT with and without wake rotation.

Blade Element Theory

The momentum theories are based on a consideration of the mean axial and rotational velocity in the slipstream and determine the thrust and torque of a blade from the rate of decrease of momentum of the fluid. The theories determine an upper limit to the power coefficient of any blade, depending on the free-stream wind velocity and on the power extracted, but they restrict the understanding of the effect of rotor geometry (i.e. blade airfoil section, chord and twist). The blade-element theory is an alternative method of analyzing the behavior of blades due to their motion through air.

As shown in figure, for this analysis, it is assumed that the blade is divided into N sections (or elements) and the aerodynamic force acting on each blade element can be estimated as the force on suitable airfoil characteristics of the same cross-section adopted for the blade elements. Finally assuming that the behavior of each element is not affected by the adjacent elements of the same blade, the force on the whole blade can be derived by adding the contributions of all the elements along the blade.

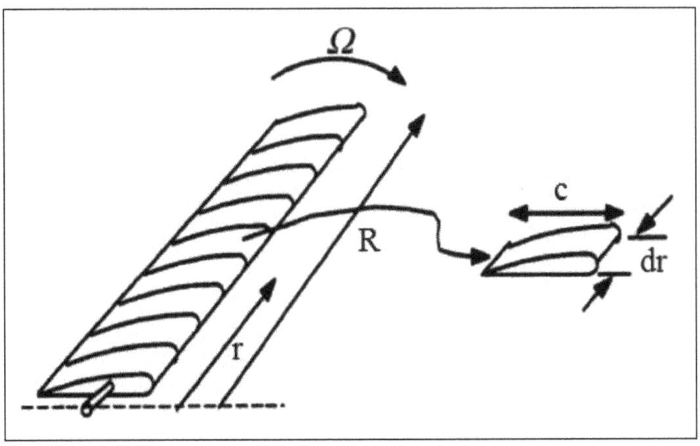

Schematic of blade elements.

A diagram showing the developed blade element at radius r and the velocities and forces acting on this element is given in figure. The relative wind velocity u_{rel} is the vector sum of the wind velocity at the rotor $u_\infty (1 - a)$ (the vector sum of the free-stream wind velocity, u_∞ and the induced axial velocity $-au_\infty$) and the wind velocity due to rotation of the blade. And this rotational component is the vector sum of the blade section velocity, Ωr and the induced angular velocity $a'\Omega r$. Hence the relative wind velocity will be as shown on the velocity diagram in figure. The minus sign in the term $u_\infty(1 - a)$ is due to the retardation of flow while the air approaching the rotor and the plus sign in the term $\Omega r(1+a')$ as shown in figure is due to the flow of air in the reverse direction of blade rotation after air particles hit the blades and so give torque.

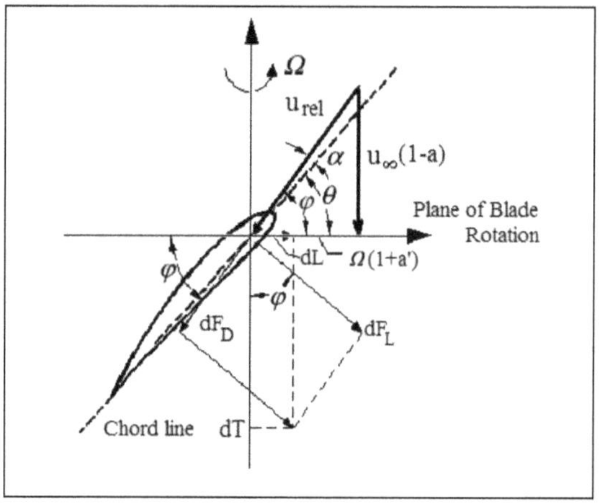

Blade geometry for analysis of a HAWT.

From the velocity diagram in figure, the following relationships can be determined:

$$\varphi = \theta + \alpha$$

$$u_{rel} = \frac{u_\infty(1-a)}{\sin(\varphi)}$$

$$\tan(\varphi) = \frac{u_\infty(1-a)}{\Omega r(1+a')} = \frac{(1-a)}{(1+a')\lambda_r}$$

$$dF_D = C_D \frac{1}{2}\rho u_{rel}{}^2 cdr$$

$$dF_L = C_L \frac{1}{2}\rho u_{rel}{}^2 cdr$$

$$dT = dF_L \cos(\varphi) + dF_D \sin(\varphi)$$

$$dL = dF_L \sin(\varphi) - dF_D \sin(\varphi)$$

If the rotor has B number of blades, the total normal (trust) and tangential force on the element at a distance r by rearranging equation.

$$dT = dF_L \cos(\varphi) + dF_D \sin(\varphi) \text{ and } dL = dF_L \sin(\varphi) - dF_D \sin(\varphi)$$

with the use of equation $u_{rel} = \frac{u_\infty(1-a)}{\sin(\varphi)}$, $dF_D = C_D \frac{1}{2}\rho u_{rel}{}^2 cdr$ and $dF_L = C_L \frac{1}{2}\rho u_{rel}{}^2 cdr$:

$$dT = B\frac{1}{2}\rho u_{rel}^2 \left(C_L \cos(\varphi) + C_D \sin(\varphi)\right) cdr$$

$$dT = B\frac{1}{2}\rho u_{rel}^2 \left(C_L \sin(\varphi) - C_D \cos(\varphi)\right) cdr$$

The elemental torque due to the tangential forces, dL operating at a distance r from the center is given by:

dQ = rdL

Hence the elemental torque by inserting equation:

$$dT = B\frac{1}{2}\rho u_{rel}^2\left(C_L \sin(\varphi) - C_D \cos(\varphi)\right)cdr$$

into equation dQ = rdL:

$$dQ = B\frac{1}{2}\rho u_{rel}^2\left(C_L \sin(\varphi) - C_D \cos(\varphi)\right)crdr$$

And by defining solidity ratio, σ as following;

$$\sigma = \frac{Bc}{2\pi r}$$

and inserting equations $u_{rel} = \dfrac{u_\infty(1-a)}{\sin(\varphi)}$ and $\sigma = \dfrac{Bc}{2\pi r}$ into:

$$dT = B\frac{1}{2}\rho u_{rel}^2\left(C_L \cos(\varphi) + C_D \sin(\varphi)\right)cdr$$

and

$$dQ = B\frac{1}{2}\rho u_{rel}^2\left(C_L \sin(\varphi) - C_D \cos(\varphi)\right)crdr,$$

the general form of elemental thrust and torque) equations become:

$$dT = \sigma\pi p\frac{u_\infty^{\ 2}(1-a)^2}{\sin^2(\phi)}\left(C_L \cos(\phi) + C_D \sin(\phi)\right)rdr$$

$$dQ = \sigma\pi p\frac{u_\infty^{\ 2}(1-a)^2}{\sin^2(\phi)}\left(C_L \sin(\phi) - C_D \cos(\phi)\right)r^2dr$$

Thus, from blade element theory, two equations $dT = \sigma\pi p\dfrac{u_\infty^{\ 2}(1-a)^2}{\sin^2(\phi)}(C_L \cos(\phi) +$ $C_D \sin(\phi))rdr$ and $dQ = \sigma\pi p\dfrac{u_\infty^{\ 2}(1-a)^2}{\sin^2(\phi)}\left(C_L \sin(\phi) - C_D \cos(\phi)\right)r^2dr$ have been obtained. They define the normal force (thrust) and the tangential force (torque) on an annular rotor section as a function of the flow angles at the blades and airfoil characteristics. Here it is convenient to consider in turn the following assumptions which are the bases of developing blade element theory:

- The assumption that the behavior of an element is not affected by the adjacent elements of the same blade.
- The airfoil characteristics to be adopted for the element.

Blade Element-momentum Theory

Blade Element-momentum (BEM) theory refers to the determination of a wind turbine blade performance by combining the equations of general momentum theory and blade element theory. In this case by equating the elemental thrust force equations from general momentum theory and blade element theory (Equations $dT = 4a(1-a)\frac{1}{2}\rho u_\infty^2 2\pi r dr$ and $dT = \sigma\pi p \frac{u_\infty^2 (1-a)^2}{\sin^2(\phi)}(C_L \cos(\phi) + C_D \sin(\phi)) r dr$ respectively) the following relationship is obtained:

$$\frac{a}{(1-a)} = (\sigma C_L)\frac{\cos(\phi)}{4\sin^2(\phi)}\left[1+(C_D/C_L)\tan(\phi)\right]$$

and equating the elemental torque derived in both general momentum theory and blade element theory (equations $dQ = 4a'(1-a)\frac{1}{2}u_\infty \Omega r^2 2\pi r dr$ and

$dQ = \sigma\pi p \frac{u_\infty^2 (1-a)^2}{\sin^2(\phi)}(C_L \sin(\phi) - C_D \cos(\phi)) r^2 dr$ respectively):

$$\frac{a'}{(1-a)} = \frac{(\sigma C_L)}{4\lambda_r \sin(\phi)}\left[1-(C_D/C_L)\cot(\phi)\right]$$

Equation $\frac{a'}{(1-a)} = \frac{(\sigma C_L)}{4\lambda_r \sin(\phi)}\left[1-(C_D/C_L)\cot(\phi)\right]$ can be rearranged by using equation $\tan(\varphi) = \frac{u_\infty (1-a)}{\Omega r(1+a')} = \frac{(1-a)}{(1+a')\lambda_r}$, which relates a, a', φ and λ_r based on the geometric considerations:

$$\frac{a'}{(1+a)} = \frac{(\sigma C_L)}{4\cos(\phi)}\left[1-(C_D/C_L)\cot(\phi)\right]$$

In the calculation of induction factors a and a', accepted practice is to set C_D zero for the purpose of determining induction factors independently from airfoil characteristics. For airfoils with low drag coefficient, this simplification introduces negligible errors. So equation:

$$\frac{a}{(1-a)}=(\sigma C_L)\frac{\cos(\phi)}{4\sin^2(\phi)}\left[1+(C_D/C_L)\tan(\phi)\right],$$

$$\frac{a'}{(1-a)}=\frac{(\sigma C_L)}{4\lambda_r\sin(\phi)}\left[1-(C_D/C_L)\cot(\phi)\right]\text{and}\frac{a'}{(1+a')}=\frac{(\sigma C_L)}{4\cos(\phi)}\left[1-(C_D/C_L)\cot(\phi)\right]$$

can be rewritten considering $C_D = 0$:

$$\frac{a}{(1-a)}=(\sigma C_L)\frac{\cos(\varphi)}{4\sin^2(\varphi)}$$

$$\frac{a'}{(1-a)}=\frac{(\sigma C_L)}{4\lambda_r\sin(\varphi)}$$

$$\frac{a'}{(1+a)}=\frac{(\sigma C_L)}{4\cos(\varphi)}$$

By using these three equations the following useful relationships result after some algebraic manipulation;

$$C_L=\frac{4\sin(\varphi)}{\sigma}\frac{(\cos(\varphi)-\sin(\varphi))}{(\sin(\varphi)+\lambda_r\cos(\varphi))}$$

$$a=\frac{1}{\left[1+\left[4\sin^2(\varphi)/(\sigma C_L)\cos(\varphi)\right]\right]}$$

$$a'=\frac{1}{\left[\left[4\cos(\varphi)/(\sigma C_L)-1\right]\right]}$$

$$a/a'=\lambda_r/\tan(\varphi).$$

Solution Methods

Two solution methods will be proposed using these equations to determine the flow conditions and forces at each blade section. The first one uses the measured airfoil characteristics and the BEM equations to solve directly for C_L and a. This method can be solved numerically, but it also lends itself to a graphical solution that clearly shows the flow conditions at the blade and the existence of multiple solutions. The second solution is an iterative numerical approach that is most easily extended for flow conditions with large axial induction factors.

The first method focuses in solving for C_L and α. Since $\varphi = \theta + \alpha$, for a given blade geometry and operating conditions, there are two unknowns in equation $C_L = \frac{4\sin(\varphi)}{\sigma}\frac{(\cos(\varphi)-\sin(\varphi))}{(\sin(\varphi)+\lambda_r\cos(\varphi))}$, C_L and α at each section. In order to find these values, one can use the empirical C_L vs. α curves for the chosen airfoil. One then finds the C_L and α from the empirical data that satisfy equation $C_L = \frac{4\sin(\varphi)}{\sigma}\frac{(\cos(\varphi)-\sin(\varphi))}{(\sin(\varphi)+\lambda_r\cos(\varphi))}$. This can be done either numerically or graphically. Once C_L and α have been found, *a'* and a can be determined from any two of equations $\frac{a}{(1-a)} = (\sigma C_L)\frac{\cos(\varphi)}{4\sin^2(\varphi)}$ through $a/a' = \lambda_r/\tan(\varphi)$. It should be verified that the axial induction factor at the intersection point of the curves is less than 0.5 to ensure that the result is valid.

The other equivalent solution method starts with guesses for *a* and *a'*, from which flow conditions and new induction factors are calculated. Specifically:

- Guess values of *a* and *a'*.
- Calculate the angle of the relative wind (equation $\tan(\varphi) = \frac{u_\infty(1-a)}{\Omega r(1+a')} = \frac{(1-a)}{(1+a')\lambda_r}$).
- Calculate the angle of attack from equation $\varphi = \theta + \alpha$ and then C_L and C_D.
- Update *a* and *a'* from equations $\frac{a}{(1-a)} = (\sigma C_L)\frac{\cos(\varphi)}{4\sin^2(\varphi)}$ and

$\frac{a'}{(1-a)} = \frac{(\sigma C_L)}{4\lambda_r\sin(\varphi)}$ or $a = \frac{1}{[1+[4\sin^2(\varphi)/(\sigma C_L)\cos(\varphi)]]}$ and

$a' = \frac{1}{[[4\cos(\varphi)/(\sigma C_L)-1]]}$.

The process is then repeated until the newly calculated induction factors are within some acceptable tolerance of the previous ones. This method is especially useful for highly loaded rotor conditions.

Calculation of Power Coefficient

The overall rotor power coefficient may be calculated using the expression for the elemental power from equation dP = ΩdQ, elemental torque from equation

$$dQ = \sigma \pi p \frac{u_\infty^{2}(1-a)^2}{\sin^2(\phi)}\left(C_L \sin(\phi) - C_D \cos(\phi)\right) r^2 dr$$

together with equation $dC_P = \dfrac{dP}{1/2\rho A u_\infty^3}$ as:

$$C_P = \frac{8}{\lambda^2}\int_{\lambda_h}^{\lambda} \lambda_r^{3} a'(1-a)\left[1-(C_D/C_L)\cot(\varphi)\right] d\lambda_r$$

Finally by using equations $\dfrac{a}{(1-a)} = (\sigma C_L)\dfrac{\cos(\varphi)}{4\sin^2(\varphi)}$ and $a = a' = \lambda_r / \tan(\varphi)$ in equation $C_P = \dfrac{2}{\lambda^2}\int \sigma C_L \dfrac{(1-a)}{\sin(\phi)}\left[1-(C_L/C_D)\cot(\phi)\right]\lambda_r^2 d\lambda_r$, the general form of power coefficient expression can be obtained as:

$$C_P = \frac{8}{\lambda^2}\int_{\lambda_h}^{\lambda} \lambda_r^{3} a'(1-a)\left[1-(C_D/C_L)\cot(\varphi)\right] d\lambda_r$$

where λ_h is the local speed ratio at the hub. Note that when, $C_D \approx 0$, the equation above for Cp is the same as the one derived from the general momentum theory (equation $C_P = \dfrac{8}{\lambda^2}\int_0^{\lambda} a'(1-a)\lambda_r^3 d\lambda_r$). An alternative expression for the power coefficient can be derived after performing the tedious algebra and by inserting equations $C_L = \dfrac{4\sin(\varphi)}{\sigma}\dfrac{(\cos(\varphi)-\sin(\varphi))}{(\sin(\varphi)+\lambda_r\cos(\varphi))}$ and $a = \dfrac{1}{\left[1+\left[4\sin^2(\varphi)/(\sigma C_L)\cos(\varphi)\right]\right]}$ into equation $C_P = \dfrac{2}{\lambda^2}\int \sigma C_L \dfrac{(1-a)}{\sin(\phi)}\left[1-(C_L/C_D)\cot(\phi)\right]\lambda_r^2 d\lambda_r$;

$$C_p = \frac{8}{\lambda^2}\int_{\lambda_h}^{\lambda}\left[\begin{array}{l}\sin^2(\phi)(\cos(\phi)-\lambda_r\sin(\phi))(\sin(\phi))\\ +\lambda_r\cos(\phi)\left[1-(C_D/C_L)\cot(\phi)\lambda_r^2\right]\end{array}\right] d\lambda_r$$

Note that even though the axial and angular induction factors were determined assuming the C_D = 0, the drag is included in the power coefficient calculation. By the same token, the thrust coefficient C_T can be found beginning from the definition of thrust coefficient in the equation as following:

$$C_T = \frac{4}{\lambda^2}\int_{\lambda_h}^{\lambda}\begin{bmatrix}\sin^2(\phi)(\cos(\phi)-\lambda_r\sin(\phi))(\sin(\phi)) \\ +\lambda_r\cos(\phi)\left[1+(C_D/C_L)\cot(\phi)\lambda_r^2\right]\end{bmatrix}d\lambda_r.$$

Hawt Blade Design

In this chapter, the application of BEM theory on the HAWT blade design and analyzing the aerodynamic performance of a rotor will be explained. The concept of tip-loss factor, the flow states in which HAWTs are operating and the introduction of airfoil selection criteria in HAWT blade design will be discussed. After giving all these necessary knowledge for a blade design, the blade design procedure for an optimum rotor and power performance prediction procedure is given. It should be noted here that various methods for HAWT blade design and predicting performance of a rotor have been studied.

The Tip-loss Factor

There was an assumption; the rotor has an infinite number of blades which was used in all theories discussed in the previous chapter. With the aid of this assumption radial velocity of the flow across the rotor plane and in the wake has been always neglected and by so the derivations of governing equations have been established. But near the boundary of the slipstream the air tends to flow around the edges of the blade tips and acquires an important radial velocity. Because the pressure on the suction side of a blade is lower than on the pressure side which causes the air to flow around the tip from the lower to upper surface, reducing lift and hence power production near the tip.

A number of methods have been suggested for including the effect of tip losses. An approximate method of estimating the effect of this radial flow and hence including the effect of tip losses has been given by L.Prandtl. The expression obtained by Prandtl for tip-loss factor is given as:

$$F=(2/\pi)\cos^{-1}\left[\exp\left[\frac{-(B/2)\left[1-(r/2)\right]}{(r/R)\sin\phi}\right]\right]$$

The application of this equation for the losses at the blade tips is to provide an approximate correction to the system of equations ($dT = 4a'(1+a')\frac{1}{2}\rho\Omega^2 r^2 2\pi r dr$,

$dQ = 4a'(1-a)\frac{1}{2}u_\infty \Omega r^2 2\pi r dr$, $\frac{a}{(1-a)} = (\sigma C_L)\frac{\cos(\varphi)}{4\sin^2(\varphi)}$, $a' = \frac{1}{\left[\left[4\cos(\varphi)/(\sigma C_L)-1\right]\right]}$

and $C_p = \frac{8}{\lambda_h}\int_{\lambda_h}^{\lambda}\begin{bmatrix}\sin^2(\phi)(\cos(\phi)-\lambda_r\sin(\phi))(\sin(\phi)) \\ +\lambda_r\cos(\phi)\left[1-(C_D/C_L)\cot(\phi)\lambda_r^2\right]\end{bmatrix}d\lambda_r)$

Thus, to order of approximation of the analysis the correct form of the axial momentum equation (equation $dT = 4a'(1+a')\frac{1}{2}\rho\Omega^2 r^2 2\pi r dr$) will be taken to be $C_T = \frac{T}{0.5\rho u_\infty{}^2 A}$:

$$dT = 4Fa(1-a)\frac{1}{2}\rho u_\infty^2\, 2\pi r dr$$

Similarly, also the angular momentum equation (equation $dQ = 4a'(1-a)\frac{1}{2}u_\infty\Omega r^2$ $2\pi r dr$) will be assumed to be:

$$dQ = 4Fa'(1-a)\frac{1}{2}u_\infty\,\Omega r^2 2\pi r dr$$

Thus the effect of the tip-loss is to reduce slightly the thrust and torque contributed by the elements near the tips of the blades.

Equations derived are all based on the definition of the forces used in blade element theory and remain unchanged. When the forces from the general momentum theory are set equal using the method of BEM theory as performed before, the derivation of the flow condition is changed.

Carrying the tip-loss factor through the calculations, the changes will be as following:

$$\frac{a}{(1-a)} = (\sigma c_L)\frac{\cos(\varphi)}{4F\sin^2(\varphi)}$$

$$\frac{a'}{(1-a)} = \frac{(\sigma c_L)}{4\lambda_r F\sin^2(\varphi)}$$

$$\frac{a'}{(1+a')} = \frac{(\sigma c_L)}{4F\cos(\varphi)}$$

$$C_L = \frac{4F\sin(\varphi)}{\sigma}\frac{(\cos(\varphi)-\sin(\varphi))}{(\sin(\varphi)+\lambda_r\cos(\varphi))}$$

$$a = \frac{1}{\left[1+\left[4F\sin^2(\varphi)/(\sigma c_L)\cos(\varphi)\right]\right]}$$

$$a' = \frac{1}{\left[\left[4F\cos(\varphi)/(\sigma c_L)\right]-1\right]}$$

$$C_P = \frac{8}{\lambda^2}\int_{\lambda_h}^{\lambda}\left[\begin{array}{l} F\sin^2(\phi)\left(\cos(\phi)-\lambda_r\sin(\phi)\right)(\sin(\phi) \\ +\lambda_r\cos(\phi))\left[1-(C_D/C_L)\cot(\phi)\right]\lambda_r^2 \end{array}\right]d\lambda_r$$

HAWT Flow States

Measured wind turbine performance closely approximates the results of BEM theory at low values of the axial induction factors but general momentum theory is no longer valid at axial induction factors greater that 0.5, because according to the equation $u_R = u_\infty(1-a)$, the wind velocity in the far wake would be negative. In practice, as the axial induction factor increases above 0.5, the flow patterns through the wind turbine become much more complex than those predicted by the general momentum theory. The thrust coefficient represented by the equation $T = 2\rho Aa(1-a)u_\infty^{\ 2}$ can be used to characterize the different flow states of a rotor. Figure below shows flow states and thrust force vectors T associated with a wide range of axial induction factors.

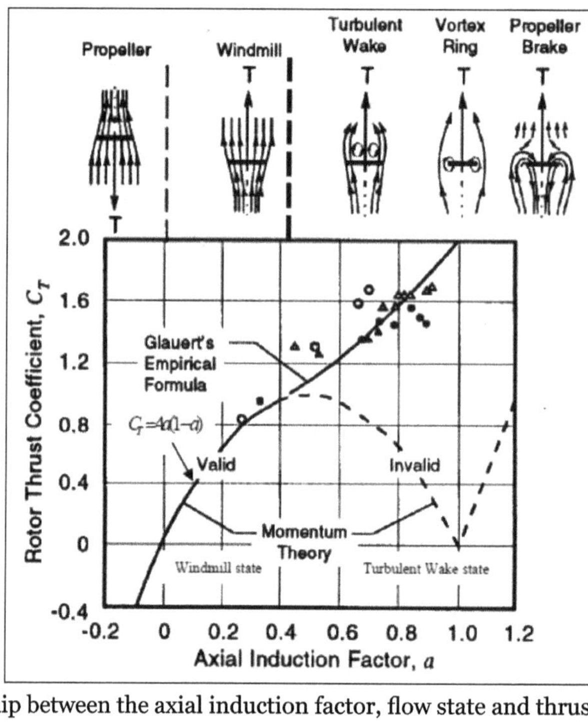

Relationship between the axial induction factor, flow state and thrust of a rotor.

According to figure above for negative induction factors (a < 0) it is simple to continue the analysis to show that the device will act as a propeller producing an upwind force (i.e. CT $C_T < 0$)that adds energy to the wake. This is typical of the propeller state. The operating states relevant to HAWTs are designated by the windmill state and the turbulent wake state. The windmill state is the normal operating state. The windmill state is characterized by the flow conditions described by general momentum theory for axial induction factors less than about 0.5. As illustrated by the data in figure above. obtained on wind turbines, above a=0.5, rotor thrust increases up to 2 with increasing induction factor in the turbulent wake state, instead of decreasing as predicted by the equation $u_R = u_\infty (1-a)$. While general momentum theory no longer describes the turbine behavior, Glauert's empirical formula for axial induction factor from 0.4 to 1.0 are often used in HAWT rotor design for predicting wind turbine flow states.

When the induction somewhat over unity, the flow regime is called the vortex ring state and when a >2.0 the rotor reverses the direction of flow which is termed propeller brake state with power being added to flow to create downwind thrust on the rotor.

In the turbulent wake state, as stated before, a solution can be found by using the Glauert empirical relationship between the axial induction factor, a and the thrust coefficient, C_T in conjunction with the blade element theory. The empirical relationship developed by Glauert, including tip losses is,

$$a = (1/F)\left[0.143 + \sqrt{0.0203 - 0.6427(0.88 - C_T)}\right]$$

This equation is valid for a > 0.4 or equivalently for $C_T < 0.96$. The Glauert empirical relationship was determined for the overall thrust coefficient for a rotor. It is customary, however, to assume that it applies equally to equivalently local thrust coefficients for each blade section. The local thrust coefficient C_{T_r} can be defined for each annular rotor section as:

$$C_{T_r} = \frac{dT}{1/2 \rho u_\infty{}^2 2\pi dr}$$

From the equation $dT = \sigma \pi p \frac{u_\infty{}^2 (1-a)^2}{\sin^2(\phi)} (C_L \cos(\phi) + C_D \sin(\phi)) r dr$ for the elemental thrust force from blade element theory together with equation $\left(C_{T_r} = \frac{dT}{1/2 \rho u_\infty{}^2 2\pi dr} \right)$, the local thrust coefficient becomes:

$$C_{Tr} = \sigma(1-a)\ (C_L \cos\phi + C_D \sin\phi) \sin\ \phi.$$

Airfoil Selection in HAWT Blade Design

Designing HAWT blade depends on knowledge of the properties of airfoils. The most significant flow factor influencing the behavior of airfoils is that of viscosity which is characterized by the Reynolds number of the airfoil/fluid combination. The Reynolds number Re is defined by:

$$Re = \frac{u_{rel} c}{\nu}$$

Airfoils in use on modern wind turbines range in representative chord size from about 0.3 m on a smallscale turbine over 2 m on a megawatt-scale rotor. Tip speeds typically range from approximately 45 to 90 m/s. Then for HAWT airfoils Reynolds number range from about 0.5 million to 10 million. This implies that turbine airfoils generally operate beyond sensitive. It should be noted that there are significant differences in airfoil behavior at different Reynolds numbers. For that reason it must be made sure that appropriate Reynolds number data are available for the blade design.

The lift and drag characteristics of airfoils show also significant aspect-ratio dependence at angles of attack larger than 30 deg. But in the fully-attached region in which HAWT is operating under windmill state are not greatly affected by aspect-ratio, so that two-dimensional (i.e. infinite aspect-ratio) data can be used in blade design at low angles of attack. However, when twodimensional data are used, tip-loss factor must be added as described in equation

$$F = (2/\pi)\cos^{-1}\left[\exp\left[\frac{-(B/2)[1-(r/2)]}{(r/R)\sin\phi}\right]\right].$$

There are evidently many engineering requirements into the selection of a wind turbine airfoil. These include primary requirements related to aerodynamic performance, structural, strength and stiffness, manufacturability and maintainability.

The usual assumption is that high lift and low drag are desirable for an airfoil and that the drag-to-lift ratio γ which is known as glide ratio as given below is a critical consideration $C_T = 4a\ (1 - a)$:

$$\gamma = \frac{C_D}{C_L}$$

Airfoils for HAWTs are often designed to be used at low angles of attack, where lift coefficients are fairly high and drag coefficients are fairly low(i.e. fairly low glide ratio).

Blade Design Procedure

Designing a blade shape from a known airfoil type for an optimum rotor means determining the blade shape parameters; chord length distribution and twist distribution along the blade length for a certain tipspeed ratio at which the power coefficient of the rotor is maximum.

As it can be seen from the equation,

$$C_p = \frac{8}{\lambda^2} \int_{\lambda_h}^{\lambda} \begin{bmatrix} \sin^2(\phi)(\cos(\phi) - \lambda_r \sin(\phi))(\sin(\phi)) \\ +\lambda_r \cos(\phi)\left[1 + (C_D / C_L)\cot(\phi)\lambda_r^2\right] \end{bmatrix} d\lambda_r$$

the overall power coefficient, C_p depends on the relative wind angle (φ), local tip-speed ratio λ_r, the glide ratio (C_D/C_L) and the tip loss factor (F). To get maximum C_P value from this equation is only possible to make the elemental power coefficient maximum for each blade element. In other words, the term in the integral of the mentioned equation should be maximum for each blade element in order to get maximum overall power coefficient from the summation of each.

As it can be seen from the equation,

$$C_P = \frac{8}{\lambda^2} \int_{\lambda_h}^{\lambda} \begin{bmatrix} \mathrm{F}\sin^2(\phi)(\cos(\phi) - \lambda_r \sin(\phi))(\sin(\phi) \\ +\lambda_r \cos(\phi))\left[1 - (C_D / C_L)\cot(\phi)\right]\lambda_r^2 \end{bmatrix} d\lambda_r$$

the overall power coefficient, C_p depends on the relative wind angle (φ), local tip-speed ratio λ_r , the glide ratio (C_D/C_L) and the tip loss factor (F). To get maximum C_P value from this equation is only possible to make the elemental power coefficient maximum for each blade element. In other words, the term in the integral of the mentioned equation should be maximum for each blade element in order to get maximum overall power coefficient from the summation of each.

For a minimum glide ratio of selected airfoil type a relationship can be established between the relative wind angle and local tip-speed ratio to determine the optimum relative wind angle, φ_{opt} for a certain local tipspeed ratio. The condition is given as:

$$\phi_{opt} \rightarrow MAX \begin{Bmatrix} \mathrm{F}\sin^2(\phi)(\cos\phi - \lambda_r \sin\phi)(\sin\phi \\ +\lambda_r \cos\phi)\left[1 - (C_D / C_L)\cot\phi\right] \end{Bmatrix}$$

When the optimum relative wind angle values are plotted with respect to the corresponding local tip-speed ratio values at which the elemental power coefficient is maximum for a wide range of glide ratios, the relationship has be found to be nearly independent of glide ratio and tip-loss factor. Hence the condition given before can be rearranged as:

$$\phi_{opt} \rightarrow MAX\left\{ \sin^2\phi(\cos\phi - \lambda_r \sin\phi)(\sin\phi + \lambda_r \cos\phi)\right\}$$

Therefore a general relationship can be found between optimum relative wind angle and local tipspeed ratio which will be applicable for any airfoil type by taking the partial derivative of the term above,

$$\frac{\partial}{\partial \phi}\left\{\sin^2\phi(\cos\phi-\lambda_r\sin\phi)(\sin\phi+\lambda_r\cos\phi)\right\}$$

Equation $\frac{\partial}{\partial \phi}\left\{\sin^2\phi(\cos\phi-\lambda_r\sin\phi)(\sin\phi+\lambda_r\cos\phi)\right\}$ reveals after some algebra:

$$\phi_{\text{opt}}=(2/3)\tan^{-1}(1/\lambda_r)$$

The procedure of blade design begins with dividing the blade length into N elements. The local tip-speed ratio for each blade element can then be calculated with the use of equation $\lambda_r=\Omega r/u_\infty=\lambda r/R$ as given below:

$$\lambda_{r,i}=\lambda(\mathrm{r_i/R})$$

Then according to equation $\varphi_{\text{opt}}=(2/3)\tan^{-1}(1/\lambda_r)$ the optimum relative wind angle for each blade element is determined as:

$$\varphi_{\text{opt,i}}=(2/3)\tan^{-1}(1/\lambda_{\text{r,i}})$$

From the equation $F=(2/\pi)\cos^{-1}\left[\exp\left[\frac{-(B/2)[1-(r/2)]}{(r/R)\sin\phi}\right]\right]$ the tip loss factor for each blade element can be found as:

$$F_i=(2/\pi)\cos^{-1}\left[\exp\left[\frac{-(B/2)[1-r_i/2]}{(r_i/R)\sin\phi_{opt,i}}\right]\right]$$

The chord-length distribution can then be calculated for each blade element by using equations $C_L=\frac{4F\sin(\varphi)}{\sigma}\frac{(\cos(\varphi)-\sin(\varphi))}{(\sin(\varphi)+\lambda_r\cos(\varphi))}$ and $\sigma=\frac{Bc}{2\pi r}$ as:

$$c_i=\frac{8\pi r_i F_i\sin\phi_{opt,i}}{BC_{L,design}}\frac{(\cos\phi_{opt,i}-\lambda_{r,i}\sin\phi_{opt,i})}{(\sin\phi_{opt,i}+\lambda_{r,i}\cos\phi_{opt,i})}$$

where, $CL_{,design}$, is chosen such that the glide ratio is minimum at each blade element.

The solidity can be calculated as follows:

$$\sigma_i=\frac{\mathrm{Bc_i}}{2\pi\mathrm{r_i}}$$

The twist distribution can be determined from the following equation which can be easily derived from the velocity diagram:

$$\theta_i = \varphi_{\text{opt,i}} - \alpha_{\text{design}}$$

where, $\alpha_{\text{design},}$ is again the design angle of attack at which $CL,_{\text{design}}$ is obtained.

Finally, the power coefficient is determined using a sum approximating the integral in equation $C_P = \frac{8}{\lambda^2}\int_{\lambda_h}^{\lambda}\begin{bmatrix}F\sin^2(\phi)(\cos(\phi)-\lambda_r\sin(\phi))(\sin(\phi) \\ +\lambda_r\cos(\phi))[1-(C_D/C_L)\cot(\phi)]\lambda_r^2\end{bmatrix}d\lambda_r$ as:

$$C_P = \sum_{i=1}^{N}\begin{bmatrix}\left(\frac{8\Delta\lambda_r}{\lambda^2}\right)F_i\sin^2\phi_{opt,i}\left(\cos\phi_{opt,i}-\lambda_{r,i}\sin\phi_{opt,i}\right)\ldots \\ \left(\sin\phi_{opt,i}+\lambda_{r,i}\cos\phi_{opt,i}\right) \\ \left[1-(C_D/C_L)\cot\phi_{opt,i}\right]\lambda_{r,i}^2 \\ \end{bmatrix}$$

As a result, for a selected airfoil type and for a specified tip-speed ratio and blade length (i.e. rotor radius), the blade shape can be designed for optimum rotor and from the calculation of power coefficient the maximum power that can be extracted from the wind can then be found for any average wind velocity.

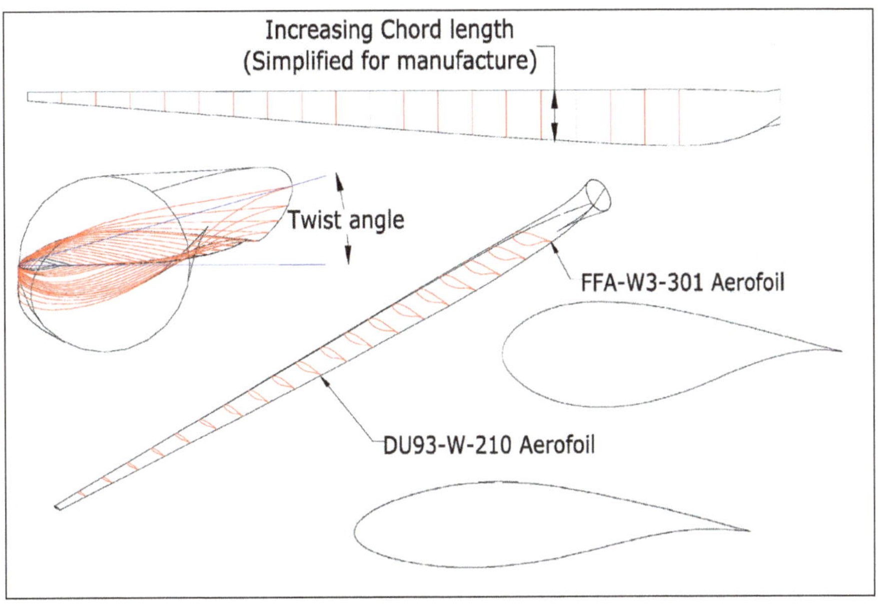

HAWT blade design.

During the design process, from the required power and wind velocity, the rotor radius, R can be estimated by including the effect of probable C_p and efficiencies. Then, choose the design tip speed ratio and the number of blades according to the type of application of the wind turbine and then select the design aerodynamic conditions, $C_{L,design}$ and αdesign, such that $C_{D,design}/C_{L,design}$ is at a minimum for each blade section. Finaly, divide the blade into N elements and use the optimum rotor theory to estimate the shape of the ith blade with a midpoint radius of r_i and equations above.

Aerodynamic Modeling and Testing of Vertical-axis Wind Turbine

Vertical axis wind turbines are of two type mainly-lift and drag models11. In this the rotating axis is maintained in vertical position i.e. perpendicular to the ground. Other mechanical and electrical components were placed almost closure to ground i.e. below the turbine using shaft. As they are not placed parallel to the ground, they have high wind catching capability from all the directions. Power and other parameters of the VAWTs can be calculated using the following equations. The kinetic energy present in the wind is treated as the power available in the wind and this is supposed to be proportional to air density, speed of the wind and the swept area of the wind turbine. Power available in the wind is given by Pw, i.e.

$$Pw=1/2*\rho*A*V^3$$

Where,

ρ is the air density i.e. 1.225 kg/m³;

A is the turbine swept area;

V is the wind velocity or wind speed.

According to a German physicist, Albert Betz, no wind turbine can convert kinetic energy of wind to mechanical energy not more than 59.3%. This conversion limitation thereafter named as Betz limit. Hence the mechanical power i.e. generated at the turbine is given as power available at turbine, Pt:

$$Pt = 1/2*\rho*16/27*A*V^3$$

The constant 16/27= ~ 0.593 referred as Betz coefficient. Electrical energy generated from the wind turbine system can be considered 70 % of 59.3 % of the input wind energy.

Therefore the coefficient of wind turbine or power coefficient, Cp, can be defined as the ratio of electricity produced by wind turbine to the total power available in the wind.

Hence Cp is given as follows:

Cp= (Power produced by the wind turbine)/ (Total power available in the wind)

Conversion to electricity would 70 % of 59.3 %, i.e. Cp value can be calculated as 0.70 * 0.593= ~ 0.41 i.e. this would determine the approximate efficiency of good turbines as 41 %. In general the efficiency range of a good quality make turbines would be in the range of 35 % - 45 %.

Swept area of the VAWT is given as the product of height of the rotor to the diameter of the rotor.

$$A = H \times D$$

where,

H is the height of the VAWT rotor in meters;

D is the diameter of the VAWT rotor in meters.

Tip speed ratio (TSR) of the wind turbine is defined as the ratio of speed of the rotating blade tip to the speed of free wind stream.

$$TSR = (\omega * R)/V$$

TSR can also be denoted with a parameter, Lambda.

Solidity of a wind turbine is used to 'define the geometry of VAWT, and it is calculated using following equation as follows and it is a dimension less parameter;

$$Solidity = (N * C)/D$$

where,

N is the number of blades used in designing VAWT;

C is the airfoil chord length;

D is the diameter of rotor.

Advantages over HAWT

VAWTs are having few advantages than HAWTs. Among those few are mentioned here:

- Suitable for low wind speed conditions.
- Suitable for small scale power generation such as residential applications etc.
- Can withstand to extremely bad weather conditions of snow, ice, sand etc.

- Lighter weight, lower construction cost and Installation costs.
- Does not cause any harm to the wild life.
- Capable of generating more power on annual basis when compared with HAWTs.
- Can withstand high wind velocities up to 65 m/s and also operate well in turbulent wind conditions.
- Service, repairs and maintenance can be done very easily as the generator, gear box and other major components are almost placed close to the ground.

Design of Vertical Axis Wind Turbine

Design of VAWT is described below in three stages-first is selecting the suitable airfoil, second is modelling of the airfoil and other components in software tools like solid works, and third is fabricating the model by selecting appropriate material.

Airfoil Selection

In this design, NACA 0012 airfoil is selected. The selected airfoil is symmetrical and the four digits of any NACA series airfoil defines the wing profile by the following:

- Maximum camber as percentage of chord is described by the first digit.
- The distance of maximum camber from the airfoil leading edge in terms of percent of the chord is described by the second digit.
- Maximum thickness of the airfoil as percent of the chord is described by the last two digits.

From the above three points, the selected airfoil profile is described that it has no camber and it has 12 % thickness to its chord length ratio.

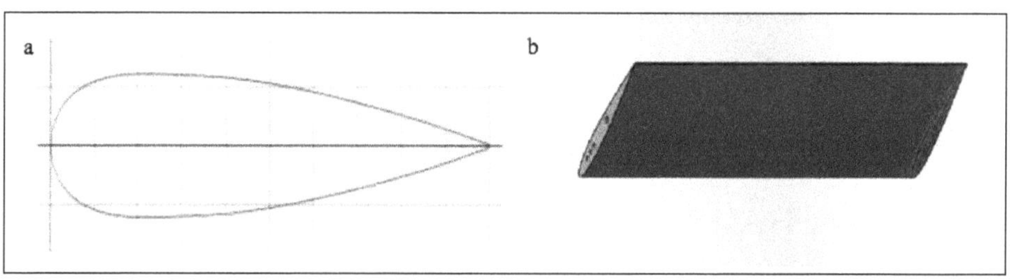

(a) NACA 0012 airfoil; (b) Straight blade profile for VAWT designed in solid works.

Modelling of VAWT

After selecting the airfoil, its co-ordinates were exported solid works software for designing the airfoil structure and to design further as per the dimension required.

The major components that are designed in this work for VAWT are airfoil shaped blades, rotor center shaft, bearing, bottom bearing holder and rotor support frames. All these components were designed individually later assembled together to make aerodynamic model of VAWT. Here straight blades are used in modelling VAWT whose radius is almost equal over its length, these is responsible for power generation over its complete length.

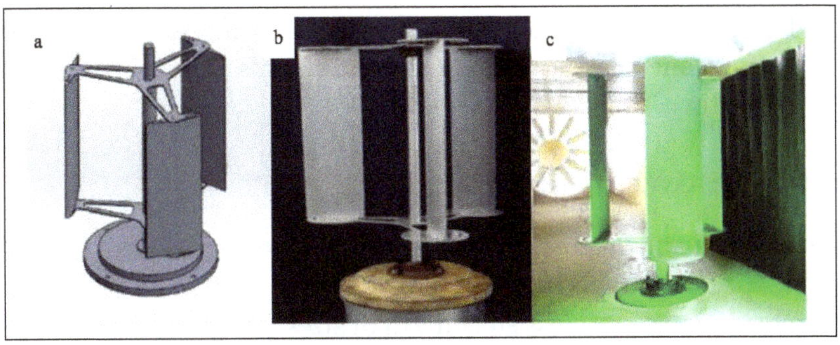

(a) Aerodynamically modelled VAWT; (b) Fabricated laboratory prototype model of VAWT; (c) Wind tunnel testing of VAWT.

Fabrication of VAWT

The experimental prototype of VAWT is designed aerodynamically using CNC machine design as per the design parameters shown in Table. The material selection for fabricating the prototype is one of the most essential requirements.

Table: Vertical axis wind turbine design parameters.

Modules or components	Parameter	Value with unit
Straight wind blades or Airfoil	Chord length Height of the blade	12 cm 40 cm
Rotor shaft	Length of the shaft Diameter of the shaft	48 cm 3.2 cm
Rotor support frame	Diameter of the frame	36 cm
Bottom base support	Diameter of support –I Diameter of support- II	30.5 cm 23.6 cm

However a variety of materials like wood, steel, aluminum, fiber materials, carbon fiber, aromatic polyamides are available for designing wind blades. Among these materials suitable one is selected based on their properties and performance. Suresh Talur et. al, presented detailed studies on material selection for wind blades, this study includes comparative analysis on giving ranking and performance percentage for different materials. Aluminum 7020 alloy is ranked 3rd whereas the other materials like stainless steel and polycarbonate sheet are in 2nd and 1st respectively. Another study on material selection by K. Suresh Babu et. al. describes which material is suitable on the basis of our requirement. In the present paper, for fabricating the laboratory testing model, aluminum material is used. Aluminum is in ductile nature having a low fatigue levels. Apart from this aluminum is having a low tensile strength, lighter weight, lower price and with good reliability. Hence aluminum is used for fabricating this laboratory testing prototype model of VAWT, shown in figure above.

Experimental Results

Any prototype model i.e. fabricated should be tested as per the facilities available for analyzing its performance or the characteristics. In this case, VAWT is tested in the wind tunnels laboratory.

Wind Tunnel Testing

VAWT is tested in the subsonic wind tunnel, whose test section size is 600*600*4000 mm. Maximum speed that can be attained in the test section is 45 m/sec with a bell mouthed entry having two stainless steel mesh of sizes 8 and 16 respectively. At the bottom of the tunnel there is a circular opening through which the base of the VWAT is fixed with the help of screws and nuts using bottom support I and II, whereas airfoil shaped wind blades, rotor supporting frames were fixed into the test section by opening top and side chambers. Once the VAWT is placed in the test section, wind tunnel is operated at different speeds. Wind speed is calculated by considering the manometers height difference using the following equation based on the Bernoulli's principle.

$$V^2 = \text{(Manometer fluid density)}/\ \text{(Air density)}*2*g*(h_2-h_1)$$

Where,

h_1 & h_2 are the height difference in the manometer because of the pressure difference in the test section of the wind tunnel; g is the gravitational force i.e. 9.8 m/sec^2;

A mathematical formulae obtained from the Bernoulli's principle gives the velocity in the test section faced by VAWT by using equation V^2= (Manometer fluid density)/(Air density)*2*g*(h_2-h_1). Now the performance of the wind turbine is analysed at different wind speeds.

Table: Experimental results of vertical axis wind turbine observed in wind tunnel testing.

Wind Speed (V)	VAWT Rotor Speed (RPM)	Power in the wind, $P_{w\,(Watts)}$	Power at the turbine, $P_{t\,(Watts)}$	Tip Speed Ratio (TSR)	Angular Velocity (w)
4.38	0	7.41	4.38	0	0
7.38	170	35.45	20.98	0.43	17.80
10.38	252	98.64	58.39	0.45	26.38
13.38	335	211.26	125.07	0.47	35.08
16.38	395	387.62	229.47	0.45	41.36

19.38	442	641.99	380.05	0.43	46.32
22.38	460	988.66	585.28	0.38	48.17

Performance Parameter Analysis

Performance parameters of VAWT is analyzed using wind tunnel testing. These parameters includes power available in the wind, power captured by the turbine, power coefficient of turbine, tip speed ratio, angular velocity and solidity of the turbine are studied experimentally and tabulated in table.

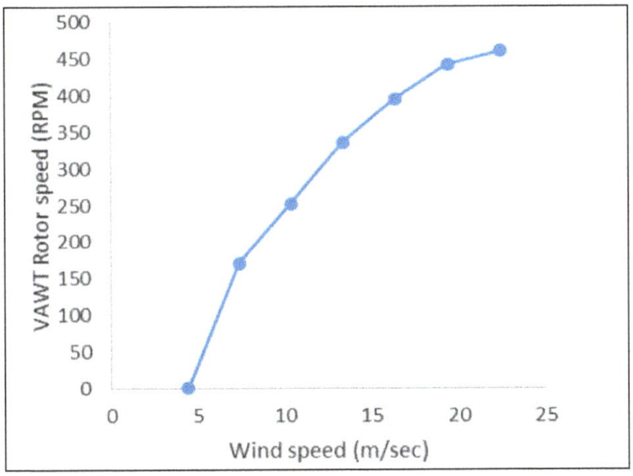

Wind speed (m/sec) vs. VAWT rotor speed (RPM).

Figure: Shows the wind speed vs. VAWT rotor speed characteristics, this plot shows that as the wind speed increases the rotational speed of the turbine rotor increases and maintains a linear relationship.

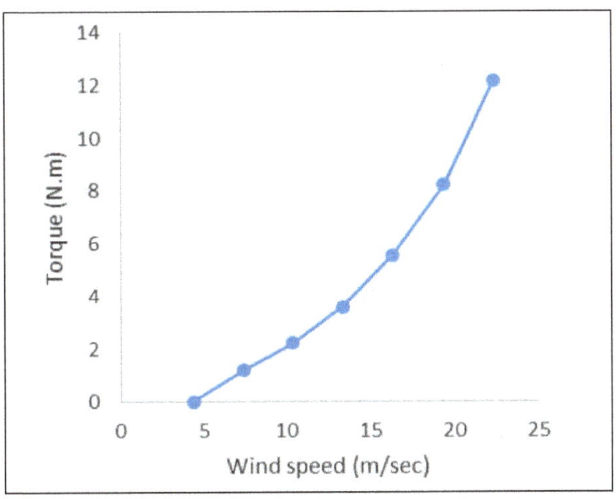

b) Wind speed (m/sec) vs. Torque (N. m).

Figure: Represents the wind speed vs. torque characteristics, this plot also shows that torque increases with increase in wind speed.

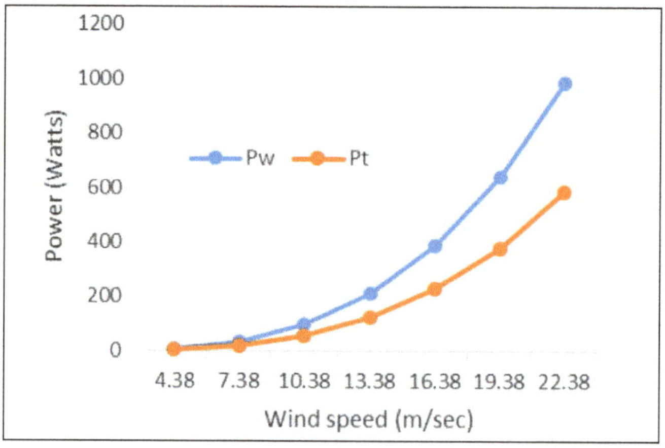

c) Wind speed (m/sec) vs. Power (Watts).

Figure: Shows the relation about power generated and power available in the wind. This practically designed and tested model of VAWT produces accountable power generation with wind speed.

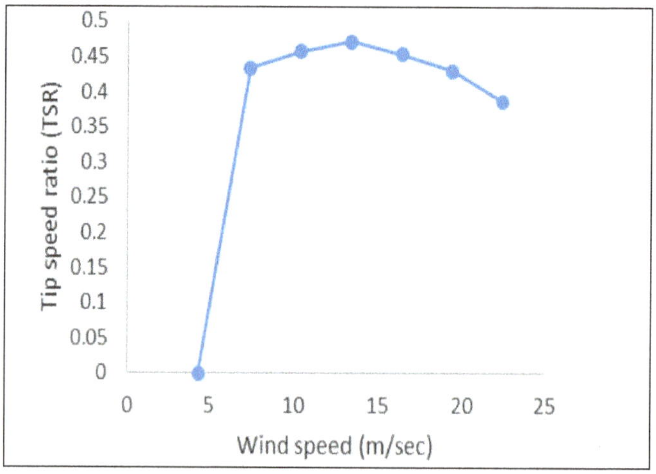

d) Wind speed (m/sec) vs. Tip speed ratio (TSR).

Figure: Represents the tip speed ratio of the VAWT in relation with wind speeds, here tip speed ratio (TSR) increases with increase in wind speed, but after reaching certain level of wind speed TSR is maintained almost a closure values. This parameter study shows the best performance of the wind turbine model.

Aerodynamic modelling of straight bladed vertical axis wind turbine model is designed. Apart from this a practical model is designed and tested in wind tunnels to test its feasibility. The analyzed experimental results evaluated the characteristics parameters of VAWT. Wind tunnel test results showed the acceptable performance. However future work should be done on selecting a high quality and lighter weight material for designing. This would help in achieving good cut in speeds. Apart from this the performance of the wind turbine can be evaluated by modelling and testing with different type of airfoils.

Measuring the Performance of a Wind Turbine

The traditional method of measuring wind turbine performance under laboratory conditions in ideal circumstances will always tend to be optimistic and rarely reflect how the turbine actually behaves in a real situation. Real performance will be affected by local wind conditions, nearby obstructions, power demand profiles and a range of other factors. There will also be deterioration in performance with time due to wear and tear. What is important is how the turbine reacts and actually delivers power on site. Without proper data this can be very subjective and a matter of personal opinion.

This application note focuses on collecting real-time power, rotor speed and wind speed data of a specific Proven WT2500 wind turbine that has been in almost continuous operation for 6 years. Information derived from these measurements can help decide whether maintenance is necessary. Further interpretation of the data could help optimise the wind turbine and could assess if it has been sited in the correct position.

Rotor Details

The Proven wind turbine is a downwind horizontal axis type. The rotor is constructed from three polypropylene blades each with two spring-loaded hinges at the blade root. This assembly is referred to as a Zebedee hinge and comprises an inner hinge (angled at 90 degrees) and an outer hinge (angled at 45 degrees). The two hinges work together to effectively twist the blade as an automatic passive blade pitch and power control system. The system is normally set to give a low angle of attack for easy starting at low wind speeds. As the turbine accelerates, centrifugal force causes the hinges to open. This flattens the blade, increasing the angle of attack to the relative airflow. The centrifugal force and therefore the pitch both increase with rotor speed. In high winds, the wind force on the blades overcomes the spring tension and the rotor disk cones, both stalling the blades and reducing the effective rotor disk area and therefore power output. An increase in shaft torque will cause the blades to oppose the action of the centrifugal force which will tend to close the hinges, decreasing the angle of attack.

The spring tension is factory-set to run at a maximum operating speed of 300 rpm (5 rev/s) at the rated 12 m/s wind speed and above. Wear at the spring mountings and loss of elasticity of the springs in use can cause stalling and coning to commence at lower wind speeds, reducing the maximum power output of the turbine.

Generator Details

The Proven wind turbine is a variable-speed three-phase permanent magnet alternator and consists of four-pole pairs rotating within 24 stator coils. Each coil nominally produces 30 volts and eight coils per phase are arranged during manufacture in series and parallel combinations to produce wind turbines generating 24, 48, 120 or 240 volts. The neutral is not accessible as a common point on production machines. The rating of 300 rpm with a four-pole-pairs generator produces a line frequency of 20 Hz.

Site Details

In the case of the wind turbine at Port Louis farm, the owner had suspected that the turbine has been under-performing and offered it for these trials.

The choice of suitable sites for small wind turbines is usually limited by the cost of the heavy–duty power cable needed to carry the turbine power output to the charge controller and battery. Most of the turbines are sited within 100 metres of the powerhouse, which contains the battery bank, diesel generator, power inverter and wind turbine charge control panel. Turbine siting is always a compromise and can result in less than ideal winds. To reduce DC power losses in the power cables, most turbines operate at 48 V (50–60 A) but a few operate at 24 V (100–120 A).

The Port Louis WT2500 wind turbine was designed to generate 2.5 kW at 48 volts in a 12 m/s wind and is 110 metres from the powerhouse on a slight rise of land to the north west of the buildings. It has been sited so that is relatively unobstructed from the prevailing westerly winds. However, being relatively low to the ground on the standard 6.5 metre mast, the turbine receives more buffeting and turbulence than if it were mounted higher.

Strategy and Design

In order to gain enough data to obtain meaningful results, three details would have to be logged in real time:

- The output power of the turbine. (watts).
- The rotation speed of the turbine (rev/s).
- The wind speed (m/s).

The analysis would require instrumentation to measure 4 parameters:

- System voltage.
- System current.

- Turbine rotational speed.
- Wind speed.

Interface and Software

A means of simultaneously measuring and recording this information in real time was devised based around the ADC-16 8-channel analogue to digital converter. The ADC-16 can measure DC signals from −2.5 to +2.5 volts DC. The PicoLog software allows scaling within the program so that correction, calibration and conversion factors can be applied automatically to the recordings on collection to minimise spreadsheet processing. Most data processing was performed using Microsoft Excel.

Additional Circuitry

Additional circuitry was required for the signal processing so that measurements could be taken for turbine rotor speed and wind speed. New signal processing circuits were designed to maximise the working range of the ADC-16 interface to improve accuracy and precision.

Both turbine rotational speed and wind speed measurements were sensed as low-frequency AC waveforms. The Proven turbine maximum rotation speed of 300 rpm with a four-pole permanent magnet alternator produces an AC frequency of 20 Hz. Similarly, the output frequency of the anemometer at 20 m/s is 25.7 Hz. To convert an incoming AC signal into a DC voltage that could be recorded by the ADC-16, two similar frequency to voltage converters were constructed based on the LM 2917 integrated circuit (IC). Component values were calculated using the manufacturer's datasheet to give an operating range of 0–25 Hz, which would be converted into an output of 0–2.5 volts to match the input range of the ADC-16.

Connecting the Test Equipment

The test equipment was connected to the Proven control panel, as shown below in figure below.

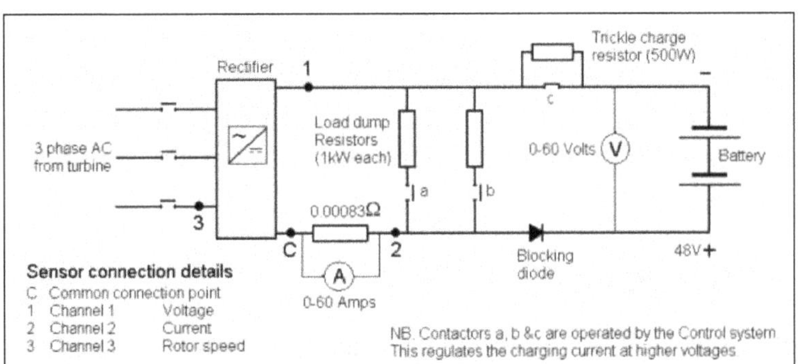

Connecting the sensors.

Channel 1 — System Voltage

A potential divider was constructed to drop the 48 volt DC line from the Proven control panel down to a measurable voltage.

Channel 2 — System Current

This was measured using the existing Proven current shunt, which was designed to give a voltage of 50 mV at a current of 60 amperes.

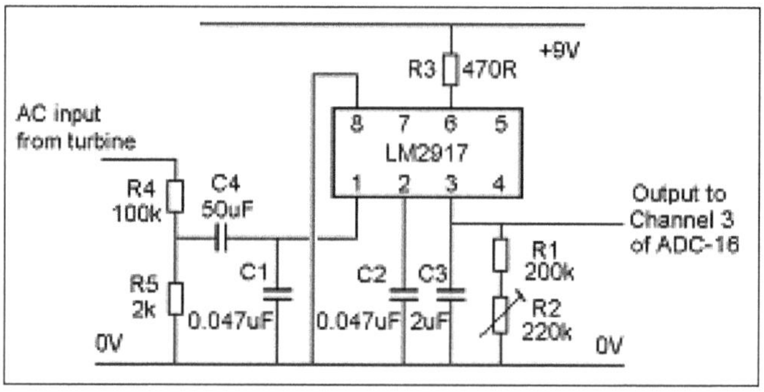

Frequency to voltage converter– Rotor speed measurement.

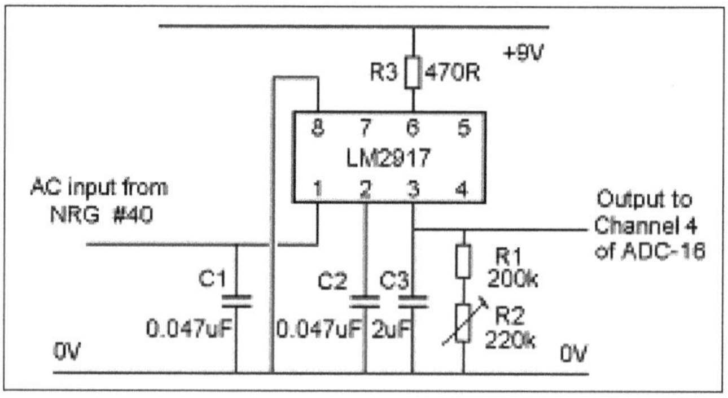

Frequency to voltage converter – Wind speed measurement.

Channel 3 — Turbine Rotational Speed

This was derived from one phase of the three-phase input to the wind turbine rectifier. A physical connection was made to one of the phases of the AC line from the wind turbine within the Proven control panel. A potential divider was designed to reduce the peak AC voltage to around 1.5 to 2 volt DC. As this was frequency measurement, the voltage was not critical. Figure shows the frequency to voltage converter that was used to create a voltage for the ADC-16.

Channel 4 — Wind Speed

The sensor used was an NRG 40 anemometer head, an industry standard unit. This was mounted at hub height about 5 metres alongside the wind turbine perpendicular to the wind axis. The NRG 40 uses a small permanent magnet generator to produce an AC signal with the frequency proportional to the wind speed. The anemometer AC output was converted to a voltage using a frequency to voltage converter.

Measurement Conditions

Ideally, the tests should have been applied to a wind turbine driving a constant impedance load. However, this was not an option as the power system was in full operation supplying a real load to the farm buildings and domestic accommodation. During the tests, the diesel generator was running supplying power and charging the battery. This helped stabilise the 48 volt DC voltage bus and minimised transient load impacts that would change load impedance and affect readings if operating on battery and wind only. The Proven dump load control system was deactivated during the tests.

The main measurements were all taken in one day when the wind steadily decreased from 13 to 6 m/s. The PicoLog software allowed some latitude in recording intervals, but typically a reading of all four parameters was taken every second and recorded in blocks of 500 seconds. As the anemometer was fixed on a lightweight portable telescopic mast, it was possible to change the position of the anemometer to maintain the relative positions if the wind changed.

Data Processing

The primary data was binned and averaged over different time blocks ranging from 5 seconds to 1 minute. This was to allow for delays in the response of the turbine due to the difference in inertia between itself and the anemometer. After examination, the data set using 30 second blocks was selected as the best option for further processing.

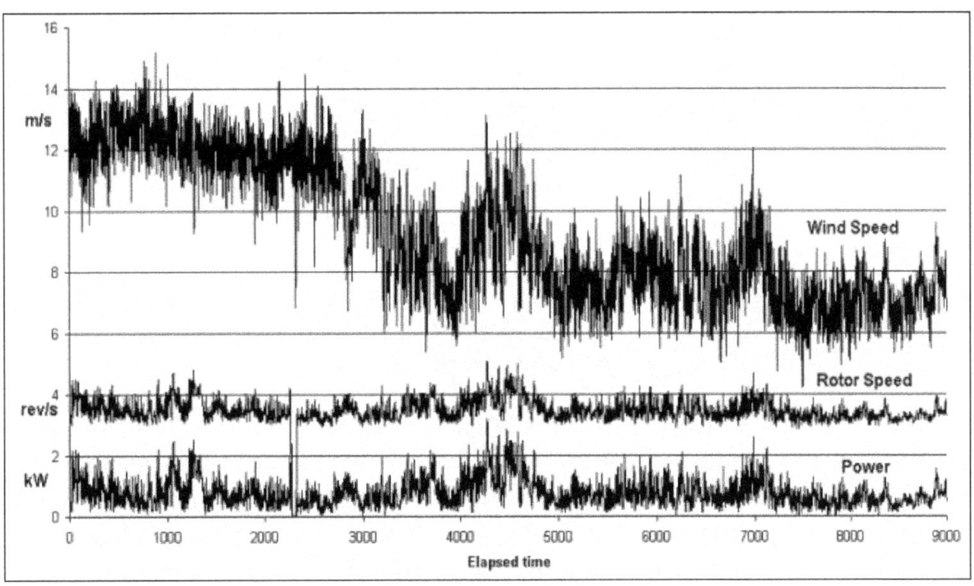

Overview of the entire test.

A power versus wind plot was plotted in Figure and the Proven WT2500 published power curve overlaid for comparison. This shows power versus wind speed for the entire run after the data had been averaged and binned in 30 second blocks.

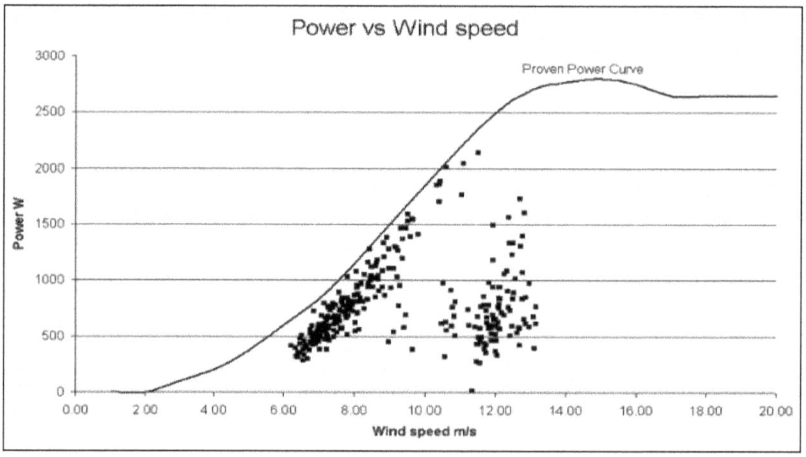

Recorded power and wind speed.

Figure compares wind speed and rotor speed for the same duration. Both graphs clearly show that there is severe over-regulation at wind speeds above 10 m/s that does not allow the wind turbine to reach its maximum power potential and it never reaches the rated speed of 5 rev/s (300 rpm).

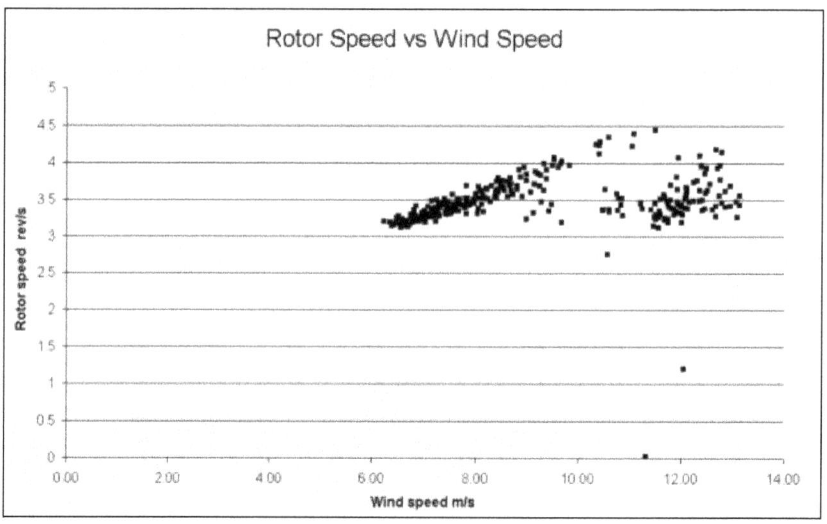

Rotor speed and wind speed.

Above 10 m/s, where power regulation caused blade coning, the early data recorded was erratic and use of these figures would distort further data processing. A cut off point was set at 3500 seconds and the readings taken before this point were not used for the detailed calculations. As the wind speed during the day did not fall below 6 m/s, and regulation cut in at 10 m/s, the final results were based on a narrow band of rotor speeds between 3.1 and 4.4 rev/s.

The universal standard to assess the performance of a wind turbine is a plot of the Power Coefficient versus the Tip Speed Ratio performance curve. This curve tells you how efficiently a turbine converts the energy in the wind to electricity. Various methods were used to plot this curve.

Field Observations

The Proven WT 2500 wind turbine is designed robustly to survive hard conditions in turbulent zones on a relatively short mast which is much nearer to the ground than other designs. As a result, it will encounter gusts and rapid but small changes in wind direction that cause it to continually accelerate, decelerate and oscillate noticeably around its yaw axis. Rarely does it run at a steady speed pointing directly into the wind.

During the tests, it was noted that the anemometer reacted very quickly to wind changes and was direction independent. On the other hand, the much greater inertia of the wind turbine would cause a delay in response in both rotation and yaw. The passive blade control mechanisms, by their nature will also lag slightly behind variations in the wind. The attitude of the blade angle due to the action of the Zebedee hinge, rotation speed and direction of yaw means that the wind turbine will rarely instantaneously match the theoretical maximum power generation values under real operating conditions. This contrasts sharply with wind tunnel tests where yaw and shear winds are not an issue and the turbine has time to settle down in between changes in loads or wind speed to present a stable set of readings reflecting its optimum performance.

High-wind Operation

It was thought that the wind turbine was underperforming prior to the test and these tests have proved this to be correct. Wind speeds in excess of 10 m/s produced an erratic power output from the turbine, well below the rated values.

Wind Turbine Tip Speed Ratio

The Tip Speed Ratio (often known as the TSR) is of vital importance in the design of wind turbine generators. If the rotor of the wind turbine turns too slowly, most of the wind will pass undisturbed through the gap between the rotor blades. Alternatively if the rotor turns too quickly, the blurring blades will appear like a solid wall to the wind. Therefore, wind turbines are designed with optimal tip speed ratios to extract as much power out of the wind as possible.

When a rotor blade passes through the air it leaves turbulence in its wake. If the next blade on the spinning rotor arrives at this point while the air is still turbulent, it will not be able to extract power efficiently from the wind. However if the rotor span a little more slowly the air hitting each turbine blade would no longer be turbulent. Therefore the tip speed ratio is also chosen so that the blades do not pass through too much turbulent air.

The tip speed ratio is given by dividing the speed of the tips of the turbine blades by the speed of the wind – for example if a 20 mph wind is blowing on a wind turbine and the tips of its blades are rotating at 80 mph, then the tip speed ration is 80/20 = 4.

Optimum Tip Speed Ratio

The optimum tip speed ratio depends on the number of blades in the wind turbine rotor. The fewer the number of blades, the faster the wind turbine rotor needs to turn to extract maximum power from the wind. A two-bladed rotor has an optimum tip speed ratio of around 6, a three-bladed rotor around 5, and a four-bladed rotor around 3.

Highly efficient aerofoil rotor blade design can increase these optimum values by as much as 25-30% increasing the speed at which the rotor turns and therefore generating more power. A well designed typical three-bladed rotor would have a tip speed ratio of around 6 to 7.

If the tip speed ratio is too low – for example if poorly designed rotor blades are used – the wind turbine will tend to slow and/or stall. If the tip speed ratio is too high the turbine will spin very fast through turbulent air, power will not be optimally extracted from the wind, and the wind turbine will be highly stressed and at risk of structural failure.

Different types of turbine have completely different optimal TSR values – for example a Darrieus wind turbine is a vertical axis (VAWT) design with aerofoil blades which generate aerodynamic lift and therefore the TSR can be high, but a Savonius wind turbine which is also a VAWT is a drag design and so the TSR will always be less than 1 – i.e. it cannot spin faster than the wind hitting it.

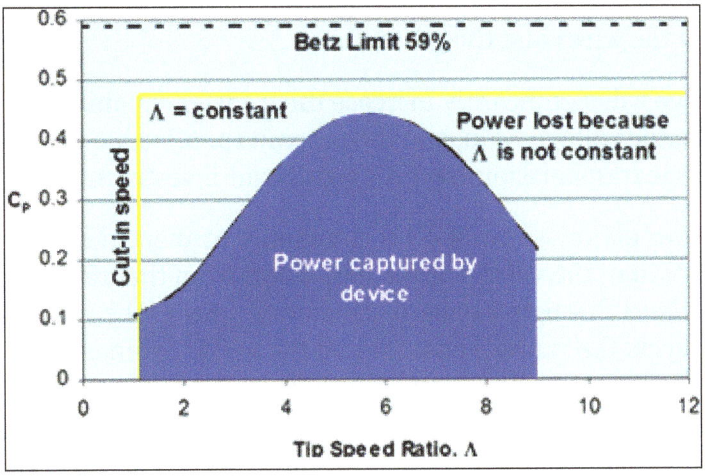

Pictured above is a graph showing the power coefficient for different values of tip speed ratio for a two-bladed rotor. The theoretical maximum efficiency of a wind turbine generator is given by the Betz Limit of around 59%. With a tip speed ratio (TSR) of just under 6, the power coefficient for this example turbine is 0.45 (= 45%).

Variable Speed Wind Turbines under Normal and Fault Conditions

To successfully deploy wind energy, we need to have the resources and the electrical energy product to the load centers. Abundant resources can be profitable only if we can generate low-cost wind energy. The transmission line, the strength of the grid, and the proximity of the wind resource to the load center are all important factors in successful wind deployment.

In an interconnected grid, the power system network and the wind power plant are interrelated. Knowing the characteristics of the wind power plant and the transmission and distribution systems is very important for identifying the problems and finding the resources to resolve the issues.

When we discuss transmission lines, we are concerned about, among other things, the stability, power transfer capability, and the losses. The stability and power transfer capability are determined partly by the impedance between the sending and the receiving ends. In the transmission lines, the resistive part of the impedance is generally much smaller than its reactance. In the sections that follow, the resistive part of the impedance is often ignored to simplify the discussion. The reactance between the two points consists of the reactance presented by the lines, the transformer, and the electric machine impedances when the line reactance is the major component of the total impedance. The line reactance is mostly determined by the line length—the geometrical distance between the wires and the ground.

Raising the voltage will significantly increase the transfer capability of the systems and reduce the losses, the voltage drop, and the per unit (p. u.) value of the reactance; however, higher voltage transmission requires significant investment.

A large wind power plant (> 100 MW) is commonly connected to a transmission line. At the main substation, the voltage is stepped down from the transmission level to the subtransmission level through the main transformer. From the main substation, a collector system collects the power generated by the wind turbines through line feeders. The collector system impedance of many turbines can be represented by its equivalence and as a single impedance.

In figure below, the single-line diagram represents a wind power plant. Each wind turbine has a pad-mounted transformer to step down the voltage from the subtransmission level to the low voltage generated by the wind turbine. In the future, as wind turbines become larger, the voltage at the point of generation may be increased to medium. The transmission line connects the wind power plant to the grid or to a very large network, which can be treated as an infinite bus. The short circuit current at the point of interconnection plays an important role in the behavior of the wind power plant.

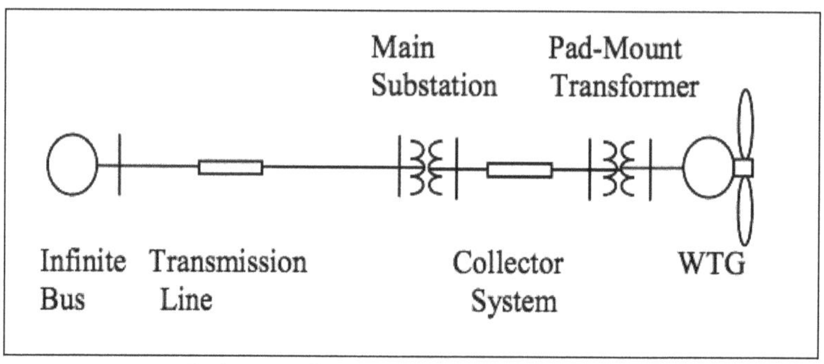

A typical network topology of a wind power plant.

Figure below shows a single line diagram that describes the nature of the fault and the network topology in pre-fault, fault, and post-fault conditions. The transmission lines consist of two parallel paths. The fault occurs at the middle of line A, and the circuit breakers disconnect line A. The remaining line B continues to deliver the output of the wind power plant. The short circuit current (SCC) of the transmission line was originally at 10 p. u.; in the post-fault condition after line A was removed, the SCC dropped to 5 p. u.

Because we are interested only in the worst-case scenario, we simulate only the weak grid condition. Let us define a prefault condition with line A and line B, which are both on line. The SCC of both lines is 10 p.u. The wind power plant consists of wind turbines with a total capacity of 100 MW. The wind turbine is generating at its full rated capacity and the wind speed is at 125% of the rated wind speed. Because of the wind speed, the blade pitch is already deployed. Initially the wind turbine operates normally, then suddenly, a three-phase fault occurs near the main substation on line A at t = 1 s. The fault lasts for nine cycles, and line A is removed at the end of the fault, which creates a weaker system with an SCC of 5.0 p.u. In this simulation, the relay protections are not set up, because we want to see the nature of the wind turbine and the system while exposed to this type of fault.

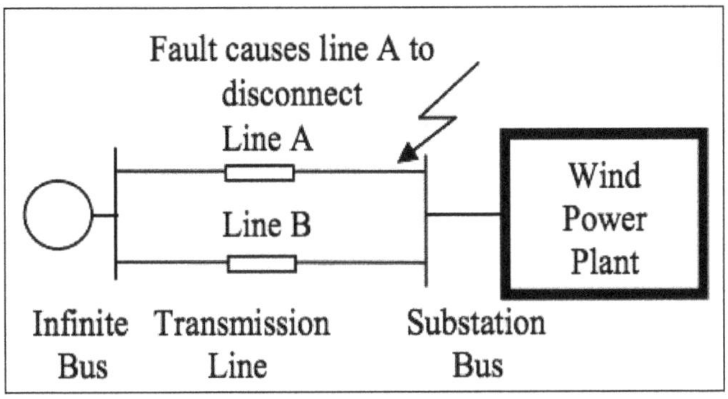

The location a fault on a weak grid.

Variable-speed Wind Turbine Generator

Although doubly fed induction generators are commonly used by the wind industry today, we use a permanentmagnet, synchronous, variable-speed generator because it represents the future of WTG capability. We assume that the WTG consists of a wind turbine connected to a generator via a gearbox or direct-drive generator. The output of the generator is connected to the utility via a power converter. The power converter provides a buffer between the utility and the generator.

Normal Operation

Under normal operation, the VSWTG operates at variable speed. At low to medium wind speeds it operates at optimum pitch angle (e.g., $\beta = 0°$). At high wind speeds, the wind turbine operates at constant speed, and the pitch angle controls the rotor speed to be constant.

The output power of the wind turbine operating in this mode is a cube function of the rotor speed until it reaches rated speed. Once the rated speed is reached, the wind turbine is operated at constant power by controlling the pitch angle. Normally, the pitch angle controller operates very smoothly, controlling the rotor speed so it does not exceed the rated speed.

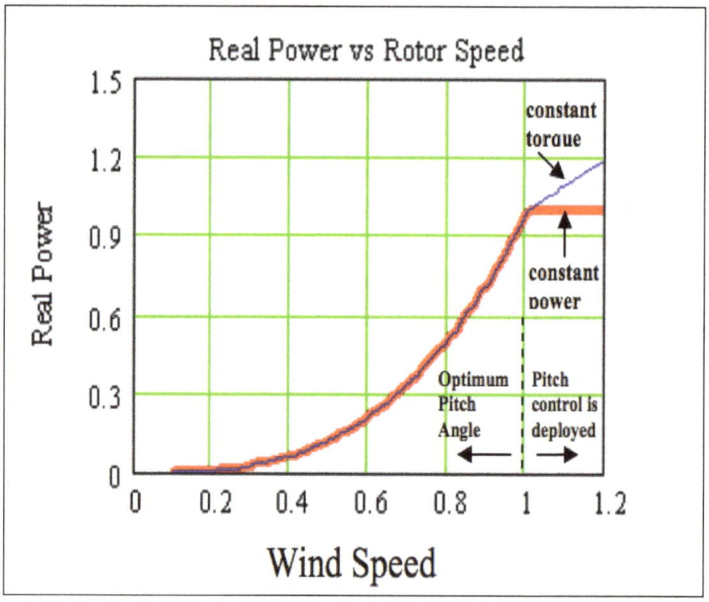

Power curve of a typical VSWTG.

This pitch control mechanism, in effect, sheds aerodynamic power to keep the rotor speed from reaching a runaway condition. The wind turbine brakes will be deployed if the blade pitch control fails.

Figure above shows the power curve of the VSWTG. Because the aerodynamic efficiency of a wind turbine is maximized at a specified tip-speed ratio, it is controlled so that it operates at that specific tip-speed ratio. This condition can be achieved when the ratio of rotor speed to wind speed is constant. Two curves are shown in Figure. Below rated speed, the curves coincide. Above rated speed, however, the curves deviate. The thick line represents a controller that maintains constant power above rated speed; thus, above the rated speed, the torque is inversely proportional to the rotor speed. The thin line represents a controller that maintains a constant torque above rated speed, so the power increases in proportion with the speed. The constant torque capability above rated speed can better prevent the wind turbine from reaching runaway condition. However, this capability requires a good oversized power converter with an extended speed range. In practice, the pitch controller will keep the rotor speed from over-speed conditions. In most cases, the blade pitch controller of a modern wind turbine accurately limits the rotor speed at or below the rated rotor speed, thus avoiding the need for a larger power converter.

From the power converter perspective, the output of a wind turbine is limited by the generator capability and, more importantly, by the power converter capability. The generator limit is normally the thermal limit (generated mostly by losses in the windings). The power converter limit is more stringent, because power electronics are used as the building blocks of the power converter. The instantaneous value of the maximum current limit and the voltage blocking capability limit should never be violated.

Thus, the power converter can be operated within the circle of current limit value. For a VSWTG, the power converter KVA limit is represented by a circle boundary, which shows that the operating point can be anywhere in all quadrants as long as the tip of the complex power S_0 is inside the boundary. Thus, we can generate or absorb real or reactive power within the circle.

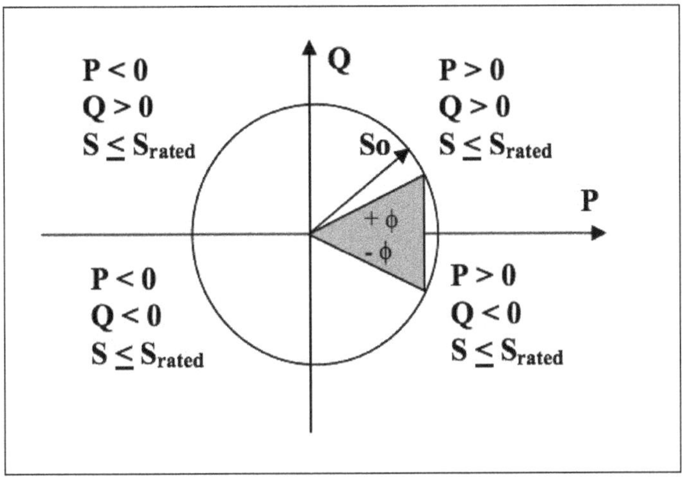

The operating boundary of a VSWTG with power converter.

Real and reactive power can be controlled independently. A modern wind turbine can usually generate rated real power with a power factor (PF) range of 0.95 leading or lagging. The shaded triangle in Figure shows the capability of the turbine. Because the terms leading and lagging power factor can be confusing in a generator or in a motor mode, the term overexcited is used to indicate that reactive power is generated by the equipment, and underexcited is used to indicate that the reactive power is absorbed by the equipment.

The VSWTG is generally operated in either PF mode or voltage control mode. With a PF controller, the wind turbine is operated at a preset PF (usually unity PF) from low to high wind speeds. The voltage is controlled by other means (capacitor, SVC, Statcom, synchronous condenser, etc.). In voltage control mode, a reference voltage is used to control the reactive power output of the generator to keep the voltage constant. In a large wind farm, two control modes may be used. Some turbines may be set at voltage control mode, while other turbines may have preset power factors.

Figure shows (a) the control diagram, (b) the phasor diagram for constant PF control, and (c) the constant voltage control. The wind turbine is assumed to be connected to an infinite bus Vs. Between the wind turbine and the infinite bus is a reactance Xs. The control block diagram (a) describes the constant voltage or constant PF control. Only the major loops and components are shown for simplicity. The VSWTG has three major components: the generator, the power converter, and the controller. Both electrical power and the electrical signal are interchanged among the generator, the power converter, and the utility. The input is the wind power feeding the wind turbine blades, and the output is the electrical power entering the utility grid. There are two choices of control objectives: PF or V. The controller has an input signal PF* or V* as the commanded power factor or the commanded reference voltage. This value is compared to the actual value PF or V. It uses the error signal to control the power converter, so that the actual PF or V moves closer to the commanded PF* or V*. The output of the controller is the reactive power command Q* that must be followed by the power converter to achieve the objective function (PF = PF* or V = V*).

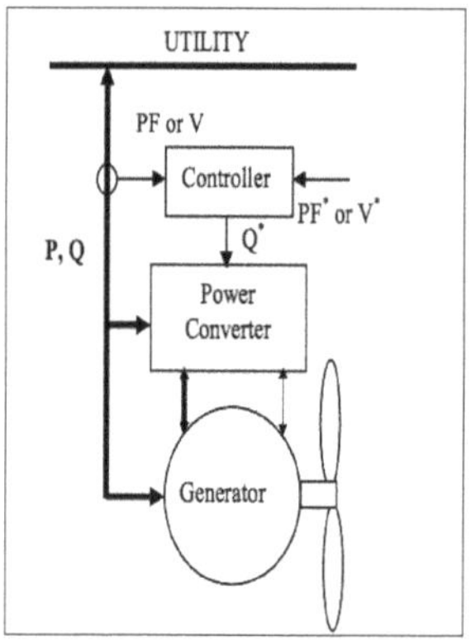

a) Control block diagram.

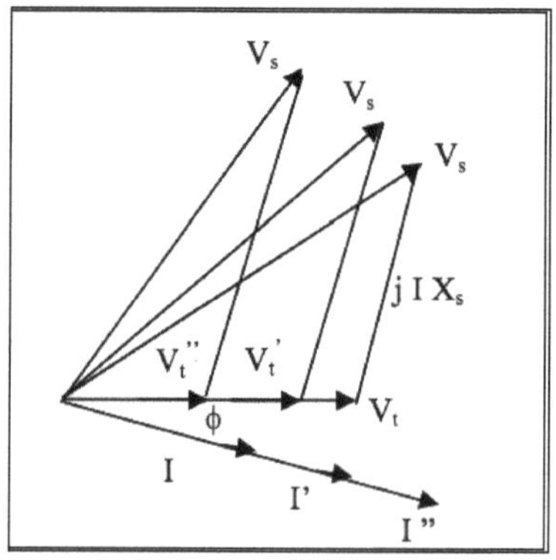

b) Constant power factor control.

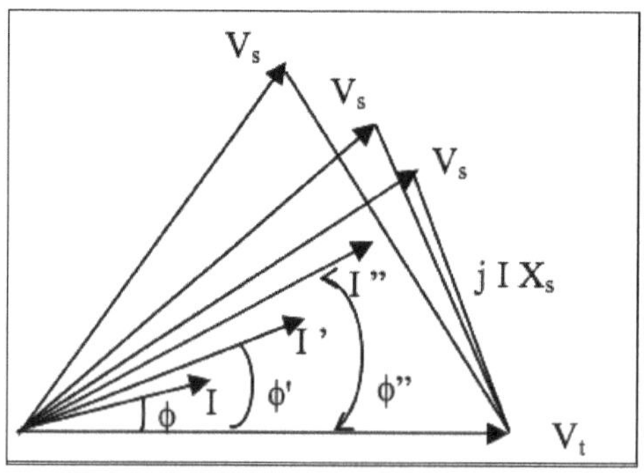

c) Constant voltage control.

Phasor and control block diagram for a VSWTG.

The output of the power converter is connected to the generator. Another loop to control the aerodynamic power is not included in this block diagram. Arrows show that the power can flow through the power converter and the generator to include the doubly fed induction generator (DFIG), where the output power can be accessed directly from the stator winding and from the rotor winding via a power converter.

Constant Power Factor

Figure shows that at constant power factor (cos φ = constant), the following can be observed:

- Current magnitude and the power angle (δ) between the wind turbine voltage V_t and the infinite bus voltage V_s increases, as the power level increases.

- Terminal voltage at the wind turbine changes with the increase of power level.
    - Underexcited condition (the generator absorbs reactive power), the terminal voltage decreases as the power level or wind speed increases.
    - Overexcited condition (the generator supplies reactive power), the terminal voltage increases as the power level or wind speed increases.

As an illustration, let us consider a wind turbine connected to an infinite bus through a line feeder. As the worst-case scenario, we want to consider an underexcited condition with PF = 0.95 with no additional reactive power compensation. Let us change the line feeder to reflect different SCCs of the lines. Figure presents a set of curves showing the terminal voltage versus output power of the wind power plant. Each curve has a different SCC. A very stiff grid (SCC = 50) performs best when the voltage varies minimally with the increase of output power. At SCC = 10, the terminal voltage drops to 0.96 p.u. at rated output power.

At SCC = 5, the terminal voltage drops to 0.9 p.u. at rated power, and at SCC = 2.5, the voltage collapses at 86% rated power. Then, short circuit capabilities will be investigated (SCC = 10, 5, and 2.5).

Figure shows the terminal voltage for a wind power plant connected to a weak grid. As the output power increases, the voltage drops differently for the three PFs. At unity PF and 0.95 PF (under an excited condition), the voltage drops accordingly. When the wind turbine is overcompensated to 0.95 (overexcited), the terminal voltage rises as the output power increases. At unity PF, the terminal voltage drops by < 2% at rated output power. In an overexcited condition, the terminal voltage rises by < 5%. The same phenomenon is also applicable to a stiff grid condition.

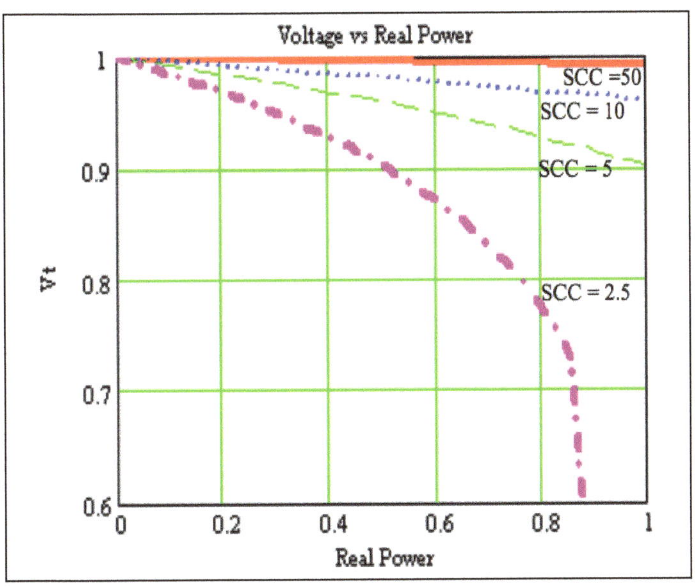

Terminal voltage as a function of real power for a VSWTG operated at PF = 0.95 under-excited.

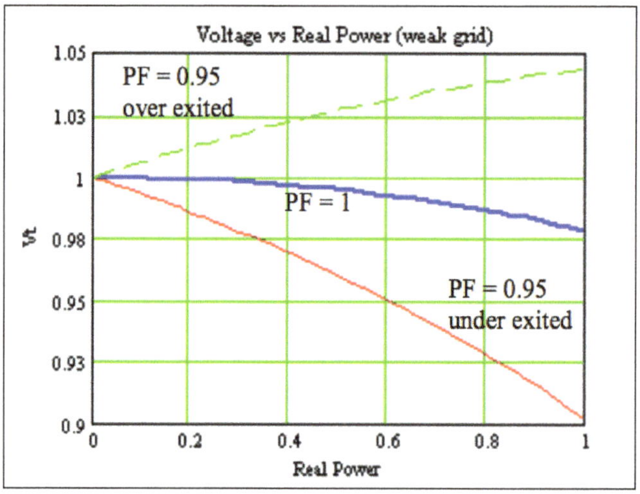

Terminal voltage versus real power for three PFs for a weak grid condition.

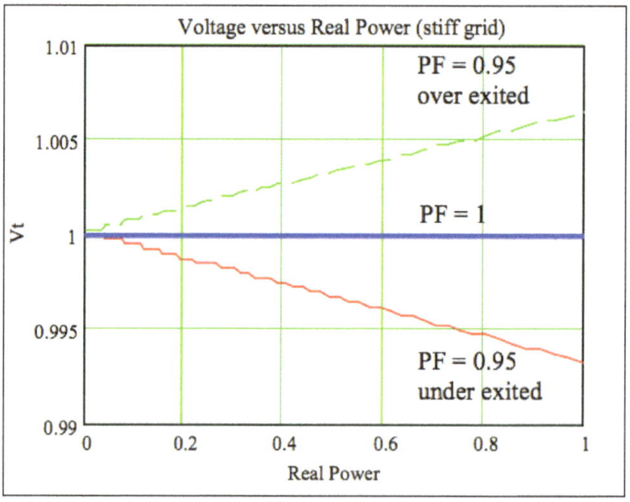

Terminal voltage versus real power for three PFs for a stiff grid condition.

The only difference is that the spread of voltage variation (from underexcited to over-excited) is much narrower ($\Delta V < \pm 1\%$) than in a weak grid.

Figure shows the terminal voltage variation as the wind turbine is connected to a stiff grid. The spread between overexcited and underexcited conditions is much narrower for a very stiff grid.

Constant Voltage Operation

The phasor diagram that describes the constant voltage control of a VSWTG is also shown in Figure. The expanded description of the voltage control concept can be found in Section II. The following can be said about constant voltage control:

- Assume that $V_s = V_t = 1.0$ p.u.

- The wind turbine must be operated in an overexcited condition.
- The current magnitude and phase angle between the wind turbine voltage V_t and the infinite bus voltage V_s increases as the power level increases.
- The phase angle (φ) between the voltage V_s and the current I, increases as the power level increases.
- The power angle (δ) between the voltage V_s and the voltage V_t, increases as the power level increases.
- $\varphi = \delta/2$.

Let us first consider different levels of grid stiffness to illustrate the concept of constant voltage control. As shown in Figure, the wind turbine must generate the reactive power to keep constant voltage at its terminal. The positive value of reactive power shown in figures is chosen to represent the reactive power generated by the wind turbine. The reactive power generated to maintain constant voltage varies with the level of grid stiffness. For example, for an SCC = 2.5, at rated output power, a wind turbine must generate reactive power as much as 21% of rated power. At SCC = 5, the reactive power needed is 10% of its rated value, and at SCC = 50, the reactive power needed is only 1% of its rated output. Thus, it is easy to see that the stiffer the grid, the less reactive power needed to keep the voltage constant.

The control strategy can also be illustrated by plotting the PF of the operation. Figure shows the PF at each operating point for constant voltage operation. For the weakest grid (SCC = 2.5), the wind turbine must be operated at PF = 0.98 (overexcited) to keep the voltage at the terminal constant.

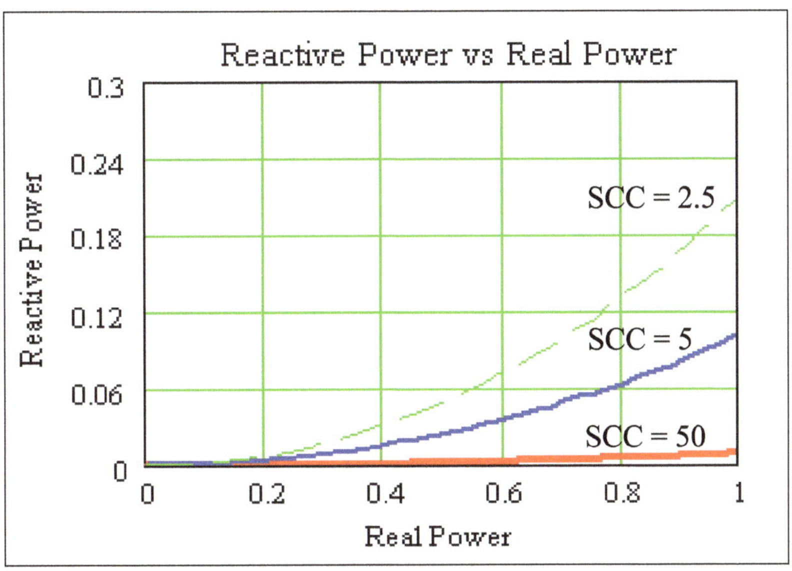

Terminal voltage versus real power for three different levels of grid stiffness.

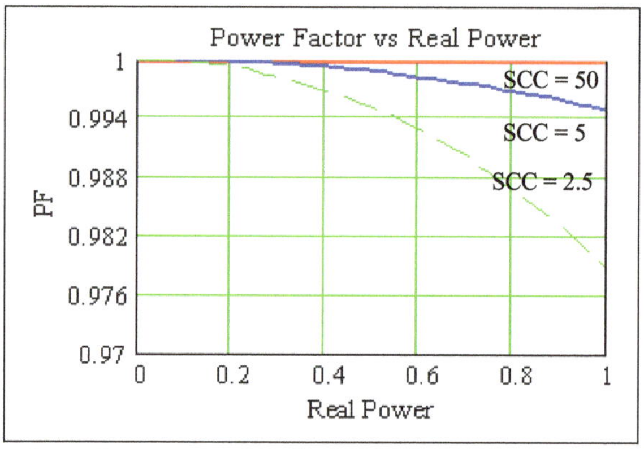

Operating PF to keep the terminal voltage constant for three levels of grid stiffness.

For a grid with an SCC = 5, the PF at rated power must be kept at 0.99 (overexcited). For a very stiff grid (SCC = 50), the wind turbine barely produces reactive power. Thus, we can conclude that the variable-speed wind turbine we studied can easily fulfill this requirement (power factor = ± 0.95, leading and lagging).

Operation under Fault

VSWTGs tend to be more forgiving in operation under a fault condition because they can vary the speed and adjust the excitation (control reactive power via power converter).

The significance of variability in speed is that the rotor of the wind turbine is a type of energy storage (kinetic energy). The kinetic energy stored is proportional to the rotational inertia and the square of rotor speed. Thus, increasing the speed to twice the original value means that the kinetic energy increment is four times higher.

The significance of adjustable excitation is that the terminal voltage can be adjusted at will (within the allowable range). Obviously, the operating point of the WTG and the severity of the fault affect the survivability of the operation. Figure shows the power curve under various fault conditions.

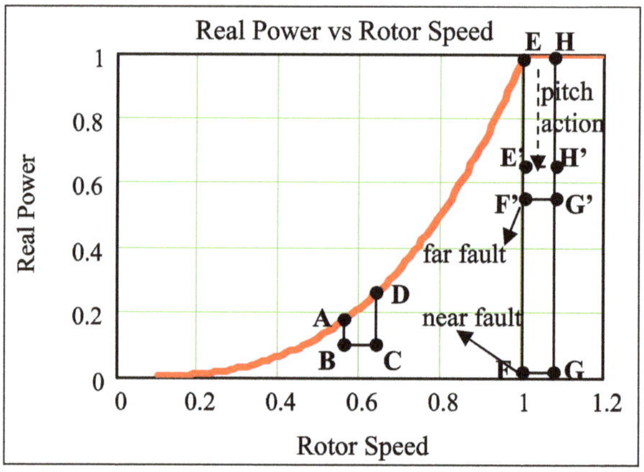

Power curve during fault conditions.

Most Favorable Condition

The most favorable condition is a far fault when the wind turbine operates in a low wind speed condition. For a far fault, the terminal voltage is relatively high (non-zero), so it is easily corrected by the power converter. In addition, the WTG can still generate sufficient real power (there is enough generator torque to oppose aerodynamic torque), so a run-away condition is less likely. In a low wind condition, the rotational speed of the turbine is relatively low, so there is enough headroom to store kinetic energy from low rotor speed to rated rotor speed. Let us take a closer look at a far fault. Referring to figure, assume that the fault occurs when the operating point of the WTG is at point A. The voltage at the terminal drops, assuming the VSWTG operates under constant voltage control, and the reactive power increases to increase the voltage to normal. If the power converter manages to control the voltage generated to, say, 50% of pre-fault value during the fault, the electrical power or the torque will be reduced to 50% of its normal value (line BC). The power converter switches have an absolute current limit that cannot be exceeded without damaging the power semiconductor. The pitch controller is normally not operated below rated speed. The aerodynamic torque remains the same as in a pre-fault condition. The aerodynamic power (line AD) accelerates the wind turbine about 30% in a pre-fault condition. The acceleration is illustrated by the polygon ABCD. As the wind turbine accelerates, the aerodynamic power has to oppose two things: (1) the electrical torque remains during the fault; and (2) the change in kinetic energy = 0.5 J $\left(\omega_B^2 - \omega_A^2\right)$. As the speed increases, the fault will probably clear in four to nine cycles. But even if the circuit breaker fails to clear the fault, the time and torque are sufficient to reduce acceleration. Let us assume that the fault condition remains uncorrected until the rotor speed reaches rated speed. At rated speed, the pitch controller starts to control the pitch angle to limit the rotor speed.

Least Favorable Condition

The least favorable condition is a near fault, and it happens during high wind speed conditions. A near fault condition indicates that the remaining voltage at the terminal is very low, so the output power will decrease significantly to accommodate the current limit of the power converter. The generator power opposing the aerodynamic power (represented by line EH) will be much lower than in a normal condition. The acceleration is represented by the rectangle EFGH. At high wind speeds, the wind turbine operates at rated speed, and the pitch angle has been deployed to reduce the aerodynamic power. It is adjusted to keep the rotor speed from exceeding rated speed. Thus, the rotor speed does not have enough headroom to increase and to change the energy acquired aerodynamically into kinetic energy. In addition, the blade pitch angle is no longer zero but at a high position close to the upper limit pitch angle. The blade pitch angle also does not have enough headroom to shed aerodynamic power. The worst case is if the pitch controller fails to shed aerodynamic power. The fault is not cleared, and the rotor runaway problem can only be remedied by mechanical brakes (usually two alternative brakes must be provided to avoid this condition).

Let us back up a few steps and say that the fault is not a near fault, but is farther away from the WTG, so the voltage does not drop to zero. The generator power is illustrated by line F'G'. Now let's say that the wind speed just reaches rated wind speed, so the blade pitch is barely deployed. If the pitch controller is fast enough to reduce the aerodynamic power, the aerodynamic power can be reduced quickly, say from line EF to line E'F'. In this condition, the acceleration is illustrated by E'F'G'H', which is much smaller than the area EFGH. Thus, even at high wind speed the possibility of a runaway condition is remote.

Dynamic Simulation

To illustrate the dynamics of a VSWTG in a fault condition, let us use a VSWTG with a permanent-magnet generator that uses a power converter to process the electrical power and return it to the utility line. This system is commonly called full-power conversion as opposed to partial power conversion, as in a DFIG.

For this particular run, the generator is set to generate rated output power at unity PF (reactive power is set to zero). Initially, the wind turbine is operated normally when the fault occurs at t = 1 s. The simulation is continued up to 10 s. The duration of the fault is nine cycles. The fault occurs in the middle of two parallel high-voltage transmission lines. The fault is cleared by disconnecting one of the two parallel lines. Before the fault, the SCC of the line is 10 (SCC = 10); after the fault is cleared the SCC is 5 (SCC = 5). The wind speed during the event is set at 125% of rated wind speed, so the pitch angle has been deployed at 13°. The upper limit of the blade pitch angle is 30°. and the optimal blade pitch angle is 0°.

Figure below shows the terminal voltage of a VSWTG operated under constant PF at PF = 1. The terminal voltage has some residual voltage during the fault at about 0.18 p.u.; at the fault site, the voltage drops to zero. Small drops occur on the voltage traces before and after the fault occurs because the operation is at constant PF, so there is no attempt to correct the voltage. The voltage drop is caused by the difference between SCC = 10 and SCC = 5 (a weaker grid) when one of the parallel lines is disconnected.

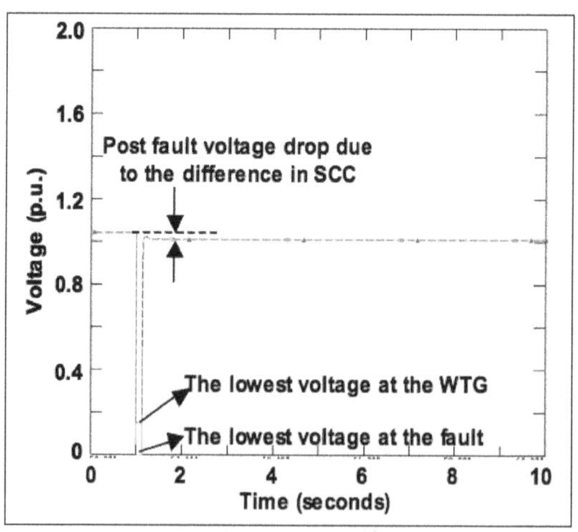

Terminal voltage for a VSWTG at PF = 1.

Figure below shows the real and reactive power output of the VSWTG. The real power drops to a very small value during the fault because the current limit of the power converter has been reached. With a low voltage during the fault, the maximum power that can be generated during the fault is limited by the current limit of the power converter. In the postfault condition, there is a small reduction in output power because the voltage is reduced. Because the control is set to provide a unity PF, the reactive power from the generator is set to zero.

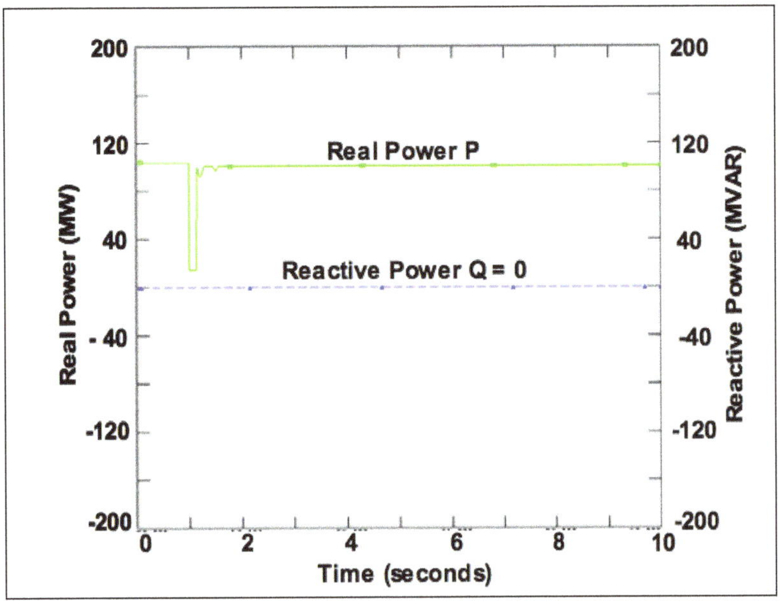

Real power and reactive power for a VSWTG at PF = 1.

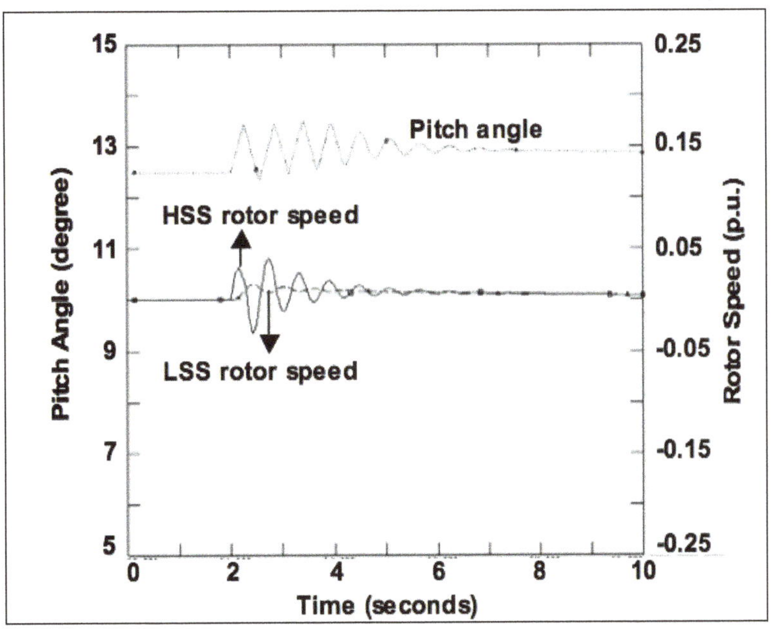

Low-speed shaft, high-speed shaft rotational speed, and pitch angle for a VSWTG at PF = 1.

Figure above shows the variation of the high-speed and lowspeed shaft rotational speeds. The spread of the speed variation is < ± 4%. The pitch angle varies between 12.5° and 13.5°. For this fault event, the wind turbine stabilized to normal (no runaway condition). In this example, the grid condition is considered weak (SCC = 5), and the duration of the nine-cycle fault is considered long (four cycles clearing time is normally used for general studies).

References

- Wind-turbine-design, wind-energy: alternative-energy-tutorials.com, Retrieved 14 July, 2019
- Working-principle-of-wind-turbine: electrical4u.com, Retrieved 11 January, 2019
- 23-wind-turbine, other: raeng.org.uk, Retrieved 18 April, 2019
- Small-wind-turbines-specification-design-and-economic-evaluation, wind-turbines-design-control-and-applications: intechopen.com, Retrieved 27 May, 2019
- Theory-of-wind-turbine: electrical4u.com Retrieved 21 July, 2019
- Measuring-the-performance-of-a-wind-turbine, application-note, library: picotech.com, Retrieved 13 July, 2019
- Wind-turbine-tip-speed-ratio, wind, wordpress: reuk.co.uk, Retrieved 24 March , 2019

Chapter 4

Applications of Wind Energy

Wind energy has a wide range of applications in wind hybrid systems, wind water pumping systems, electricity generation, wind battery charges, wind electricity generators, etc. This chapter has been carefully written to provide an easy understanding of these applications of wind energy.

Electricity Generation from Wind Power

Wind power converts the kinetic energy in wind to generate electricity or mechanical power. This is done by using a large wind turbine usually consisting of propellers; the turbine can be connected to a generator to generate electricity, or the wind used as mechanical power to perform tasks such as pumping water or grinding grain. As the wind passes the turbines it moves the blades, which spins the shaft. There are currently two different kinds of wind turbines in use, the Horizontal Axis Wind Turbines (HAWT) or the Vertical Axis Wind Turbines (VAWT). HAWT are the most common wind turbines, displaying the propeller or 'fan-style' blades, and VAWT are usually in an 'egg-beater' style.

Converting Wind to Mechanical Energy

Wind is converted by the blades of wind turbines. The blades of the wind turbines are designed in two different ways, the drag type and lift type:

- Drag type: This blade design uses the force of the wind to push the blades around. These blades have a higher torque than lift designs but with a slower rotating speed. The drag type blades were the first designs used to harness wind energy for activities such as grinding and sawing. As the rotating speed of the blades are much slower than lift type this design is usually never used for generating large scale energy.

- Lift type: Most modern HAWT use this design. Both sides of the blade has air blown across it resulting in the air taking longer to travel across the edges. In this way lower air pressure is created on the leading edge of the blade, and higher air pressure created on the tail edge. Because of this pressure difference the blade is pushed and pulled around, creating a higher rotational speed that is needed for generating electricity.

Creating Electricity from Wind

To create electricity from wind the shaft of the turbine must be connected to a generator. The generator uses the turning motion of the shaft to rotate a rotor which has oppositely charge magnets and is surrounded by copper wire loops. Electromagnetic induction is created by the rotor spinning around the inside of the core, generating electricity.

Distribution of Electricity

The electricity generated by harnessing the wind's mechanical energy must go through a transformer in order increase its voltage and make it successfully transfer across long distances. Power stations and fuse boxes receive the current and then transform it to a lower voltage that can be safely used by business and homes.

Wind Water Pumping Systems

Wind Water Pumping by windmills is possibly one of man's earliest inventions with wind energy historically being used for a wide range of applications, ranging from grinding grain to sawing wood, with many other applications as well. But there are millions of people throughout the world who do not have access to clean water for all of their daily needs. In many of these situations, water is only available from wells or aquifers, but to be usable it must first be pumped from those sources.

Stand alone wind based energy systems are an attractive solution to supplying clean electricity to off-grid consumers in remote locations allowing them to be completely independent from any oil price fluctuations. A typical wind energy stand-alone system comprises of: one or more wind energy turbines with its rated power ranging from a few watts to several kilowatts (for micro, mini and small systems) depending on the electricity demand and the available wind potential, and an appropriate energy storage device, generally a lead–acid battery storage array, that is able to guarantee x-hours of autonomy for when the wind does not blow.

But as well as supplying large quantities of electricity to charge battery banks, wind turbines can also be used to pump water. The vast majority of wind turbines built in the past have been used for non-electrical applications, and historically, wind water pumps were purely mechanical devices siting high up on top of a wooden tower pumping water for watering livestock, land drainage and irrigation, and as wind turbines do not consume water, it makes them ideal for use in dry or drought-stricken areas.

Nowadays, these mechanically driven water pumps are still a good viable option, but with advances in technology there are also a number of other possible wind energy applications that require shaft power as well. These include: wind-electric water pumps and a conventional wind water pump installed in a hybrid power system.

Conventional Wind Water Pumping Tower.

The most common type of wind operated water pumping system is completely mechanical. A typical wind water pumping system includes: the wind rotor, a tower, a mechanical pump, mechanical linkage, a well full of water (or other such water source), and piping to deliver the pumped water. Also there may be some form of water storage as well as the wind water pump. For example, a large water tank, pond, or reservoirs may be used for water storage, depending on the application.

The key points to consider in a mechanical wind water pump design includes, designing or selecting the turbine rotor and pump so that the pump size and turbine blade design as well as size, are matched to the total head. Multibladed turbines are available in various diameters ranging from 1.5m up to 5 metres, (6 to 16 feet). Water pump shaft diameters ranging from 20mm to 125mm (3/4″ to 5″) for small scale water pumping.

It is important to correctly match the rotor blades to the pump because if we were to put a large diameter water pump on a small diameter turbine over a deepish well, at low wind speed the turbine blades may not be able to produce sufficient torque in order to raise the water to the required level. This mainly due to the fact that the pump acts like a brake up to some very high wind speed where the torque becomes sufficient enough for pumping to occur.

On the other hand, if we were to put a small diameter pump on a turbine with much larger diameter blades, the pump may only supply a small fraction of the required water capacity we need when the wind resource is available. Also there is a risk of damaging the pump at high wind speeds. Then wind water pumping, by its nature, typically requires fairly high torques and low pump operating speeds.

Modern, high tip-speed ratio rotors for generating electrical power have only two or three blades. The conversion of the winds energy into hydraulic energy by a typical wind pumping system requires that the wind speed is greater than about 5-6 mph (9-10 kph), for pumping to occur. Obviously the force of the wind flowing through the blades of the turbines rotor to lift the water will depend upon the weight of the water being pumped and the speed at which the water flows.

Then the weight of the water being lifted and the speed at which the water flows will determine the power that must be delivered by the turbine to the pumping system. A deeper well means a higher head height and a heavier load of water which is why many wind water pumping systems use multibladed designs.

The vast majority of wind powered mechanically driven water pumps in operation today are of type types of design: reciprocating piston pumps and screw type rotational pumps.

For a reciprocating piston pump the turbine is connected to a gearbox and crankshaft which converts its rotary motion into reciprocating up-down motion on a pump rod which is connected to a piston in the pump at the bottom of the well pipe. When the piston is lifted by the piston rod, the piston lifts the entire volume of water above it, and the water flows out of a discharge pipe at the top. At the same time, a slight suction or vacuum is formed under the piston, causing water to flow in under the piston refilling the void.

During the next half of the cycle, the piston moves down, causing a piston valve to open, and water flows through into the top of the piston ready to be lifted again during the next half-cycle. The volume of water displaced during each stroke depends upon the piston diameter, which is the same as the diameter of the inside of the cylinder and the length of the stroke. This up-down motion of the piston pump means that the flow of water is not constant but will be inherently pulsating due to this reciprocal action.

A much simpler water pump is the spiral screw pump which is based on a long Archimedes screw type design. The main advantages of a screw pump are its simple design, reliability and slow rotation speed making it ideal for high volume heavy duty pumping. Also, driven by the motion of the turbine blades, the screw pump continually rotates producing a constant flow of water at its outlet.

The screw pump consists of a fully enclosed pipe extending down into the well and a continuous helix screw which extends for the whole or part of the length of the cylinder, thus forming a lifting chamber. As the screw rotates a small quantity of water is scooped up by the lower blades. The design the screw is such that on the next revolution this water is lifted up to the next blade, whilst the original lower blade scoops up another quantity. Then the head height of the water being lifted is that between the blades.

By placing the lower end in water and revolving the screw, water is raised to the top blade by blade. The amount of water pumped will depend upon the diameter of the cylinder and screw, the distance between blades and the speed of rotation. Therefore, the screw pumps efficiency is a function of leakage losses and blade geometry.

Wind Electric Water Pumping System

But as well as the mechanical wind water pump, it is also possible to use a wind turbine generator to pump water, (solar power can also be used to pump water). One of the main disadvantages of a mechanical wind water pump is that to pump the water it must be located above or very closed to the water reservoir. However, in the case of wind turbines that supply electrical power to pump water, the wind turbine can be placed far away from the water reservoir, in order to maximize the available wind power at that location.

For a typical wind electric water pump, the wind turbine generator used to power the system could be any conventional wind turbine generator used for stand alone off-grid applications. Instead of its generated power being used to charge batteries or power the home, the electrical power can be used directly to supply an electrically driven water pump.

Electric water pumps that are plugged into an outlet tend to use alternating current (AC) are of the centrifugal type, driven by a conventional induction motor. Also, the wind turbine generators used in home wind electric applications are typically high speed permanent magnet AC generators supplying 120 or 240 volt power to the home or utility grid.

Wind electric water pumping systems are designed to use low voltage 12 or 24 volt direct current (DC) provided by a wind turbine. Instead of using an AC motor to drive the water pump, a DC motor with a voltage and power rating compatible with the DC turbine generator could also be used allowing it to be connected directly to the pump motor. The advantage here is that by only supplying electricity to the load, high starting torque low tip-speed ratio multi-bladed rotors are not required in this type of arrangement.

Also, by using a wind electric water pumping system, pressure and/or float switches and possibly some form of microcontroller can be used to switch the water pump on or off when needed as well as for control purposes. With suitable controls, the turbine and pump would operate only when the water was needed or over a range of different pumping speeds.

Unlike a mechanical water pumping system which requires the availability of the wind to pump water, an electric wind water pump system could use batteries to store the winds energy so that water could still be supplied even during periods of low wind. A computer controlled wind electric pumping system could even be configured to run a water filtration system to clean the water for domestic use or control one or more pumps from a single wind turbine generator.

Wind turbines are well suited for harnessing the power from a renewable energy source. Wind water pumping has the advantage of being an effective low cost efficient solution to ensuring that your water is pumped in an eco-friendly way. The disadvantage of mechanically driven pumps is that they have to be sited above the water resource whereas electrically driven pumps can be sited some distance away from the turbine generator with power being supplied by an electrical cable.

The advantage of a wind electric water pumping system is that they can be configured to automatically pump and store water even during periods of no wind. Also being an electrical system, as water demand increases a wind only or a hybrid wind/solar water pumping system can be used.

Wind Hybrid System

The term wind hybrid system describes any combination of wind energy with one or more additional sources of electricity generation (e.g. biomass, solar or a generator using fossil fuels). Hybrid systems are very often used for stand-alone applications at remote sites.

The combination of renewable energy technologies allows a more balanced electricity supply during day/night and seasonal changes: At most sites wind speed is low, when the sun is shining and reaches higher values on cloudy days. Thus the amount of energy generated by wind energy reaches its maximum in the winter months, while the output of PV-cells is significantly higher in the summer. Other important examples are Wind-Diesel systems often used in remote areas. A diesel generator will be used as backup, if the electricity demand can not be covered by the installed wind turbines. Regulation and conversion of the available energy sources is a central issue planning a wind hybrid system. Many hybrid systems are uses as stand-alone off-grid applications.

Wind hybrid-systems generally consist of generating units, storage facilities and system electronic devices:

- One or more wind converters of x_o kw.
- One or more other electricity generation options either using renewable energy (RES) (e.g. Photovoltaic Panels) of fossil energy sources (e.g. Diesel generator).

- An energy storage device.
- An AC/DC rectifier of x_r kw in case the energy storage installation operates on DC current (e.g. All types of batteries).
- A charge controller of x_c kw.
- A Uninterruptible Power Supply '(UPS) of x_p kw in order to guarantee high quality AC electricity generation.
- A DC/AC inverter of x_p kw.

Energy Storage

Hybrid systems contain an energy storage device to store the surplus energy during times of high energy production, which can be used for supply when production from renewable sources is low (e.g. no wind). For this reason, the size of the device is often described by the period of time in hours h_o the average load can be covered using the storage as the sole source of energy. Other important characteristics are the overall efficiency of the storage device (determined by the loss of energy during the charge and discharge-process), the output voltage U_b and the maximum permitted discharge. Lead-acid batteries today are the most common technology solution used in hybrid energy systems. There are several alternatives like flywheels, pumped hydro storage, hydraulic storage and fuel cells.

By storing surplus energy and operating as an additional energy source, when production from the RES-source is low, the independence of the hybrid system is increased. If a diesel generator is part of the system, the storage will allow a more efficient management of the given generation units, avoiding emissions by a more efficient utilization of the diesel generator: Frequent shut-down and restart procedures as well as very unefficient generator loads can be avoided. In this manner, depence on the availability of fuel and thus on fuel price variability is reduced.

Regulation by a storage device improves the quality of the supplied power, because variations in the frequency of the current can be minimized and voltage control is available. The degree of this improvement clearly depends on the size of the storage device and the adjustment of the whole system.

It must be mentioned, that a storage device significantly raises the initial costs of the hybrid system. Desregarding the type of storage which is chosen (pumped hydro, batteries, flywheels) the environmental impacts have to be considered. Losses during charging and discharging-processes lower the efficiency of the whole system reducing the positive effect of avoided generator utilization.

System Electronic Devices

- AC/DC rectifier: In case the energy storage device consists of batteries, the three-phase AC current generated by a wind turbine has to be converted in a DC current for charging. This task is achieved by an *AC/DC rectifier* of a nominal power x_r corresponding to the rated power of the wind turbine x_o.
- DC/DC charge controller: The AC/DC rectifier connects the generating units with the *DC/DC charge controller* of a rated power of x_c charging the battery system with a charging voltage U_{cc}. Besides the charge controller distributes the incoming energy between the charging process and other DC loads which have to be covered within the hole hybrid system. This description is valid for systems using batteries as energy storage device. For a storage fed by an AC current (e.g. pumped hydro storage) the output of the generating units certainly does not have to be converted. Nevertheless in this case a controlling unit is needed for distribution of energy between storage and system loads.
- DC/AC inverter: The energy stored in the batteries has to be reconverted into AC current before it can be used to supply a load. Thus a DC/AC inverter has to be included.

Wind-Diesel Hybrid Systems

The combination of a diesel generator and a wind turbine in a hybrid system a very common and frequently used in remote areas. The following description includes considerations about system design and sizing of the components. Central questions of system designare explained by the example of the wind-diesel hybrid system.

Components

Since an equally distributed power supply by RES-sources is essential for the efficiency of a hybrid system, the *wind turbine* generator should be applicable for maximal power point tracking (MPPT). During times of low wind speed the tip speed ratio of the wind turbine must be adjusted by controlling the electromagnetic torque. Thus the wind turbine should be supplied with a modern power electronic control. For wind velocities higher than the rated wind speed of the turbine, pitch-control is an important tool to keep currents and voltages within safety limits. These control mechanisms influence the sizing of the system, because the capacity of controllers and inverters could be decreased including a lower capacity overhead needed to protect the system components against overloads.

The wind generator can either be a syncronous or an induction generator. Often used modern generator types are permanent magnet syncronous generators (PMSG) or doubly-fed induction generators (DFIG). In case of induction generator a source for excitation, either by excitation capacitators or by grid-connection must be available. In modern stand-alone hybrid-systems a direct-driven version of PMSG is the prevailing choice of wind turbine, because it does not require an external DC current for excitation. Problems to keep frequency at 50 Hz during low wind velocities are avoided by modern construction concepts with a large number of poles.

The diesel generator should be supplied with a syncronous generator. A first order modell with a single time constant can be chosen. The single time constant describes the ratio between fuel consumption and mechanical torque production.The action of the speed governor is controlled by an integral controller gain.

System Design and Sizing

Load assessment: If the planned hybrid-system is the only source of energy for a village or community, it will be important to categorize the loads which have to be supplied in the village. Medical centers are an example for high-priority loads, while economic and agricultural loads can be labeled as medium-priority loads. Finally domestic supply can be considered as a low priority load in most cases under the assumption, that electricity is a scarce resource. By categorizing the prevailing loads and collecting information about the distribution of these loads during the day the project developer creates the base for choosing the size of the system components.

Sizing of the generation units: The essential question to be answered concerns the ability of the combination wind turbine/storage to cover the high-priority loads. Designing a system which covers these loads by RES-sources or storage means the system is generally based on wind energy with the diesel generator as additional or back-up energy source. This operation strategy assures, that a great amount of fuel can be safed, which is important for gaining indepence from high fuel prices and variability.

Using the value of the high-priority load P, the basic wind energy function $P = \frac{1}{2} \cdot \left(\rho .\ A . v^{3}\right) \cdot \eta \cdot c_{p}$ can be used for calculation of the rotor diameter of the wind turbine, provided that all other variables in the function are known.

Conditions at the site must be checked: Is it possible to install such a wind turbine on this site at a height with an appropriate wind speed?

Optimization of a wind-diesel hybrid system is a complex procedure, because besides the covering of the loads the utilization of the diesel generator and the storage system must be considered. Diesel generators typically have a minimum load for operation (20%-40% of rated capacity) and their efficiency varies significantly with load level. Additionaly frequent on/off-switching of a generator causes severe wear and tear processes and should be avoided for that reason.

The storage systems consist of a battery bank, a bi-directional power electronic converter and a current limiting impedance. It can be considered as a management tool between fluctuating loads, wind energy production and the efficient utilization of the diesel engine. The charging-discharging rates of the battery technology chosen should be able to cope with these tasks in the system. Commonly used battery types are lead-acid and nickel-cadmium batteries with energy densities of the order of 0.05 kWh/litre and 0.1 kWh/litre.

Sizing of converters: Converters have a great influence on the overall system efficiency, because they have a certain amount of self-consumption and their efficiency varies with load. At two-thirds of its rated capacity the input-output efficiency of converters is typically about 87-95%, but during times of very low loads (which occur very often especially in domestic supply), the converter efficiency falls rapidly and can reach values under 50%. Thus converter size must be chosen carefully: System failure will be caused, if the converter is chosen to small and overloads occur. If the size of the converter is too big, it will operate on inefficient load levels during long time periods.

Two strategies are implemented to avoid the difficulties of converter-sizing: In certain systems the number of converters is equal to the number of loads assuring that the influence of the converter on the system performance only occurs, if the related load has to be covered. The second strategy creates to separate systems: One DC-System for lighting and other DC-applications and one AC-System, which is only used in case an AC-application has to be supplied.

Wind-PV Hybrid Systems

Wind-Hydro Hybrid Systems

One main disadvantage of any hybrid system are the high initial cost, caused by the number of necessary parts in the system and especially by the relatively high costs of energy storage systems and – in case of their integration – PV-Panels. In remote areas alternative solutions like grid extensions or dependence on sole diesel generation are expensive as well. While grid extension may result in exorbitant installation costs, pure diesel generation causes a very low initial investment but the Maintainance and Operation Costs are a very substantial factor due to fuel prices and consumption. As a result planning a wind hybrid system requires a comparison of overall costs containing initial installation costs and maintainance and operation costs.

Installation Costs of Stand-alone Hybrid Systems

Kaldelli proposes cost functions for each of the components, which include size and capacity. In this manner costs of the components have to be calculated seperatly and can finally be summarized by a simple addition. A hybrid system consisting of a wind turbine, a PV-array, a diesel generator and a battery bank is considered. The cost functions of main parts and electronic devices are presented in the following.

For a small wind turbine (rated power x_o < 100 kW) can be estimated using:

$$IC_{WT} = \left(\frac{a}{b + N_0^x}\right) \cdot N_0$$

where according Kaldelli the paramater can be chosen as follows for the european market: $a = 8,7*10^5$ €/kW; $b = 621$; $x = 2.05$; $c = 700$ €/kW. The price for PV panels can be approximated by

$$IC_{PV} = \alpha \cdot P_t \cdot x \cdot N^+.$$

where α is a function adressing the scale effects, if a high number of PV-panels x is used. For a small wind hybrid system α is near 1 decreasing with the size of the system. N^+ is the power of one PV-panel in kW_p. P_t are the cost of a PV-array expressed in €/kW_p.

The proposed function for the costs of the diesel generator proposed by Kaldelli is:

$$IC_d = \beta \cdot N_d$$

where β is the specific price of kW of capacity. Appropriate values are $\beta = 150 - 250$ €/kW.

The costs of the battery bank are a main factor in initial cost estimation of a system. For lead-acid batteries Kaldelli provides the following relationship,

$$IC_{bat} = \delta \cdot Q_{max}^{1-\omega}$$

with $\delta = 5.04$ €/Ah and $\omega = 0.078$. The size of the electronic devices is aligned differently depending on their position in the system: Inverter size is related to the capacity x_p of the UPS. Both inverter and UPS costs are included in the function. Rectifier and charge controller are align to the size of RES-Sources x_o. Thus the complete costs of electronic devices can be estimated by:

$$IC_{el} = \eta \cdot x_p^{1-\gamma} + B \cdot N_0$$

where the parameters are set to the following values by Kaldelli $\eta = 483$ €/kW, $\gamma = 0.083$, $B = 380$ €/kW. Finally Kaldelli introduces an open variable called "balance of the plant cost", which he describes as a fraction of the wind turbine cost by,

$$f \cdot IC_{WT}$$

According to Kaldelli *f* varies between 0.15 and 0.50, 0.3 is considered as a common value.

Maintainance and Operation Costs of Wind Hybrid Systems

M&O costs can be divided into variable maintainance costs VC_n and fixed M&O costs FC_n. VC_n generally represents the costs to replace system components with a lifetime r_k smaller than the lifetime of the whole system. According to Kaldelli the system VC_n can be expressed by summing up all replacement costs during the lifetime of a hybrid system:

$$VC_n = IC_0 \cdot \sum_{k=1}^{k=k_0} r_k \left(\sum_{l=1}^{l=l_k} \left(\frac{1+h_k}{1+i} \right)^{l \cdot n_k} \right)$$

where r_k is the replacement coefficient indicating the fraction of initial costs IC_0, which is caused by the replacement of the system-part k. The second term in brackets introduces the annual change in prices of the component as well as the technological improvements into the calculation. The value $1+h_k$ is a meta-variable representing these changes. By this sum function the periodical necessity of replacement of different parts of the system is represented.

Compared to this, the fixed M&O costs can be calculated in an easier manner. The fraction of FC_n given by the annual capital costs can be expressed as fraction m of the initial capital investment IC_0 corrected by a term for the annual inflation rate g_m. Given a certain annual fuel consumption, fuel costs can also be considered as fixed M&O cost. Entirely the function for FC_n is given by the following expression:

$$FC_n = m \cdot IC_0 \cdot \left(\frac{1+g_m}{1+i} + \left(\frac{1+g_m}{1+i} \right)^2 + \ldots + \left(\frac{1+g_m}{1+i} \right)^{n-1} + \left(\frac{1+g_m}{1+i} \right)^n \right)$$

$$+ c_t \cdot M_t \cdot \left(\frac{1+e_f}{1+i} + \left(\frac{1+e_f}{1+i} \right)^2 + \ldots. + \left(\frac{1+e_f}{1+i} \right)^{n-1} + \left(\frac{1+e_f}{1+i} \right)^n \right)$$

where M_f is the annual fuel consumption, c_f is the annual fuel price and e_f is the fuel price escalation rate.

Applications

T/C Stations

Stations for telecommunication in remote areas have to be supplied with power during long time periods. Extension of the electricity grid in most cases is a big financial effort, while the supply by sole diesel generators causes additional fuel- and maintainance-costs permanently. Small hybrid systems can be used to reduce fuel-consumption: A small wind turbine may be placed on the relay-mast of the T/C station, avoiding the additional installation costs of a turbine tower. As load variations of a T/C stations are rather low but a steady supply is needed, hybrid systems combining different RES-sources are preferable for this application. A battery storage for system back-up is necessary.

If the T/C station supply should be provided mainly by RES-sources, larger wind turbines on separate towers have to be installed. The inclusion of a PV-System results in reduced variations in RES-output and allows the reduction of the necessary storage size. A well designed hybrid system minimizes the fuel costs of the station supply. Several examples have shown the efficiency of fuel savings gained by the application of hybrid systems: Kaldelli refers to a system installed for remote T/C stations in Kenya consisting of a 7.5 kW turbine, sealed batteries and an inverter, which reduced fuel consumption for remote T/C stations by 70-95%.

Small Desalination Systems

Scarcity of potable water is often found in remote island situations, which at the same time often have a good wind potential. In some areas of water scarcity, water desalination contributes a considerable part to potable water supply. Implementing wind-based water desalination units in these areas is a cost-saving and emission-avoiding alternative to running the desalination units by fossil fuels. The techniques described as most efficient for water desalination, are reverse osmosis and mechanical vapour compression. Unfortunately variations and interruptions in power supply are very unfavourable conditions for water desalination plants. For this reason hybrid-systems – based on wind, but allowing a less fluctuating energy supply – combine the advantages of renewable fuel saving energy technologies with the necessary supply security for the efficient operation of small desalination plants. A very promising development of technology are floating systems, consisting of plattforms with an installed hybrid system and micro desalination plant, floating near the coast on the water surface and using the preferable offshore wind conditions.

Water Pumping

There are many areas with scarce surface water availability but sufficient water resources in appropriate depth for pumping systems. In case of poor infrastructure for water supply, wind turbines can be used as electricity source for these systems. Common aims are providing water for domestic and community supply in remote locations, as well as cattle watering and irrigation. Agriculture still is the economic activity with the greatest water consumption (almost two thirds of worlwide water consumption). The government of India promoted the implementation of wind-pumping system very intensive.

Besides a shift to Photovoltaic-systems occured because water scarcity is more often correlated with sun radiance as with appropriate wind conditions. The price-reduction has decreased the prevailing barriers of installation costs of these systems. For the same reasons as mentioned related to desalination plants hybrid-systems have advantages in comparison to the sole use of wind or PV: The greater availability of hybrid systems is an important characteristic, because a reliable water supply not at least economically essential for the growth of plants, watering of cattle and not at least the sufficient supply of the inhabitants of remote areas.

Kaldellis (2010) refers to pilot system installed in Greece consisting of a 2 kW wind turbine, 610 W_p PV-Panel and an lead-acid battery bank: The system provides 23 m³ water per day from a depth of 30 m.

Other Applications of Wind Energy

Stand-Alone Systems for Home and Business

In many places, wind power is the least-cost option for providing power to homes and businesses that are remote from an established grid. Researchers estimate that wind produces more power at less cost than diesel generators at any remote site with an average wind speed greater than about 4 meters per second. The applications for electricity in households range from operating small household appliances to refrigeration and freezing, heating, cooling, and lighting.

Table: The accompanying table gives a representative idea of the power requirements of some household appliances.

Item Description	Power, watts	Daily Energy, watt-hours/day
Incandescent Light	60	180-720
Fluorescent Light	15	45-180
Radio	5-15	20-100
Television	15-100	30-600
Refrigerator	80-500	2,000-10,000
Village Household	60-300	300-1,200

Wind turbine performance depends primarily on rotor diameter and wind speed. The accompanying table gives an estimate of a wind turbine's output, based on wind speed and rotor diameter, in watthours/day.

The amount of power that a turbine produces depends heavily on the wind speed at the turbine height. System designers must weigh improved performance of the wind turbine at higher tower heights against the increased cost and difficulty of installing higher towers.

	Wind Speed, m/s			
Roter Diameter, m	4	5	6	7
1.5	35	70	94	117

3	152	269	386	468
5	421	761	1053	1287
7	831	1522	2107	2575

Inner Mongolia Household PV/Wind Hybrid Systems Pilot Project

The Inner Mongolia Autonomous Region has been working in the past decade to provide stand-alone renewable power systems to area households. Over the past decade, more than 120,000 households have started generating electricity with 100-300 watt wind generators. By the year 2000, the New Energy Office of IMAR plans to install 25,000 remote household systems using wind, PV, and wind/PV hybrid systems and in the longer term a total of 80,000 systems throughout IMAR.

In the first phase of the cooperation in IMAR, the University of Delaware, the U.S. National Renewable Energy Laboratory, and the Inner Mongolia team completed a levelized cost analysis of rural electrification options for several counties. The following table explains the findings. For the types of systems currently being deployed for stand-alone electrical generation in rural areas of IMAR, wind generators are the least-cost option for household electricity in the four counties.

System	Output Range (kWh/yr)	Levelized cost from field analysis
Wind only	200-640	0.50-0.63
PV only	120-240	0.77-0.83
Small hybrids	400-750	0.57-0.75
Large hybrids	560-870	0.43-0.57
Diesel Gen-sets (not continuous)	660-730	0.76-0.80
Diesel Gen-sets(continuous cycle)	480-560	1.16-1.27

Systems for Community Centers, Schools and Health Clinics

A larger system can provide power to a centralized community center, health clinic, or school. A power system for a health center can enable the storage of vaccines and radio communication for emergency calls. A power system for a school can provide electricity for computers and educational television, video, and radio. Community centers often find that, in addition to the benefits of the power, such as lighting and cooling, the “waste energy” can be used to charge batteries or make ice for sale to households.

Extending the distribution lines to individual homes and creating a “mini-grid” increases the convenience of the power system to the community.

Substantial amount of diesel fuel could be saved with a control strategy and system architecture that allows shutting down the diesel generator when the wind is sufficient to carry the load, and uses short-term battery storage to reduce diesel start-ups during instantaneous lulls in the wind.

The energy requirement and equipment size calculations are similar – on a larger scale – to those for the stand-alone system. The best-designed systems will use as much power as possible directly, instead of storing it in batteries. This reduces initial cost and complexity while delivering the wind's energy in the most efficient way. Using some or all of the turbine's output to pump water, grind grain, or run other loads not dependent on utility-grade electricity reduces the need for batteries for storage of constant-frequency AC power.

Hybrid System in Atulayan Island, Philippines

Atulayan is a small isolated island in Camarines Sur, Philippines, about 3 km. off the beach of Nato, a larger island. Living on Atulayan are some 100 families deriving livelihood from fishing and seaweed farming.

The Camarines Sur IV Electric Cooperative Inc. (CASURECO IV) spearheaded the installation of the hybrid power system in Atulayan. The system was sized to supply an average of 36.5 kWh/day, with a peak load of 11.5 kW, enough to provide electricity for lighting and appliance use in 72 households (compact fluorescent bulbs, color TVs, radios, electric flat irons, and karaokes), a school (lights and a computer), streetlights, and the village playground (lighting during fiestas). Supply of electricity to a seaweed dryer supports the livelihood of the village's seaweed farmers.

With the integration of inverters and a battery bank, the village is able to enjoy grid-quality 24-hour AC power, 220V single-phase and 380V three-phase. Since the system was developed specifically for remote areas, it has automatic start-stop controls and easy operation and maintenance procedures. The villagers themselves are responsible for day-to-day management and operation of the system.

System Specifications

- 1 x Synergy S20000 wind turbine (48V) & 24 m tilt-up tower, w/ lightning protection.
- 3 x 5.5 kVA sine wave interactive inverter, tropicalized.
- 1 x 48 V battery bank 2000 Ah (C20) and battery racks c/w maintenance kit.
- 1 x Synergy 40A control box.
- 12 x 75Wp solar panels c/w pole mount array frames and wiring.
- 1 x 75 kVA gen set (retrofit of existing Atulayan unit, thus oversized).
- 1 x genset controller.
- 1 x system wiring and earthing kit plus DC panel.

(CASURECO IV provided the powerhouse, power and distribution lines.)

Industrial Applications

The number of dedicated industrial applications for wind power continues to grow. Small wind power systems are ideal for applications where storing and shipping fuel is uneconomical or impossible.

Wind power is currently being used for the following applications:

- Telecommunications.
- Radar.
- Pipeline control.
- Navigational aids.
- Cathodic protection.
- Weather stations/seismic monitoring.
- Air-traffic control.

Wind machines in industrial applications typically encounter more extreme weather than home power systems and must be designed to be robust with very minimal maintenance.

Seismic Monitoring Sites in Antarctica

In January, 1999, Northern Power Systems installed two hybrid power systems for seismic monitoring sites for on the continent of Antarctica for the United States Air Force.

The system requirements for the project were formidable: it called for 100% system reliability, a design capable of withstanding wind loads in excess of 200 miles per hour, and the interior temperature of each shelter had to be maintained at or above 50 degrees F, while exterior temperatures could reach as low as –70 degrees F.

One of the systems included dual diesel generators with a 2.2-kW solar array on the roof of the shelter. Batteries store the excess power. The other system consists of a single diesel generator, a 3.3-kW solar array, and Northern Power Systems' HR3 wind turbine. Since the sites are only accessible about 3 months out of the year, the systems required complete remote monitoring and control capability.

After one year of operation, Northern Power Systems visited the sites to perform some routine maintenance and preplanned system upgrades. Northern attests that, despite being in one of the harshest environments in the world, both systems continue to perform well.

Grid-connected Power

The cost of utility-scale wind power has been steadily declining throughout the last decade. Today, in good wind regimes, wind power can be the least-cost resource. Thanks to these positive economic trends and the fact that wind power does not produce any emissions, wind power has been the fastest-growing energy source in the world for the past few years. Wind power can help diversify a country's energy resources and can bring construction and maintenance jobs to the local community.

In large-scale wind power applications, there are two keys to developing the most cost-effective projects: wind speed and project size. Since the power output is so highly dependent on the wind speed, differences in one meter per second can mean differences of a cent or more per kWh in the cost of electricity production. Wind projects are also subject to scale economies. In general, given the same wind speed, a large project will be more cost-effective than a small one.

References

- How-does-a-wind-turbine-generate-electricity, news: goldpower.net, Retrieved 1 June, 2019
- Wind-water-pumping, energy-articles: alternative-energy-tutorials.com, Retrieved 21 March, 2019
- Hybrid-Systems-Containing-Wind-Energy: energypedia.info, Retrieved 12 April, 2019
- Purl, servlets: osti.gov, Retrieved 23 June, 2019

Chapter 5

Impacts of Wind Energy

Wind power can have adverse impacts on humans, wildlife and the environment. It includes degradation of the habitat of wildlife, plants and fish, threat to flying animals, noise pollution, visual impacts on landscape, etc. This chapter has been carefully written to provide an easy understanding of these impacts of wind power.

Environmental Impacts of Wind Energy

Operation of wind power has zero emissions of harmful substances. It does not add to global warming, the "fuel" is free, and is quite evenly distributed around the world. The energy needed to produce and install the turbine amounts to three months of turbine production. But, as with other sources of energy, wind power does have an environmental impact. The impact on wildlife is likely low compared to other forms of human and industrial activity. However, negative impacts on certain populations of sensitive species are possible, and efforts to mitigate these effects should be considered in the planning phase. Wind energy, like any other industrial activity, may cause impacts on the environment which should be analyzed and mitigated.

Environmental Benefits

What are the advantages to the environment that is caused by wind energy?

Primarily, wind energy do not cause water or air emissions, and do not produce any kind of hazardous waste as well. Moreover, wind power does not make use of natural resources like oil, gas or cause and therefore will not cause damage to the environment through resource transportation and extraction and also do not need consequent amounts of water during operation.

Wind energy is not only a favorable electricity generation technology that reduces emissions (of other pollutants as well as CO_2, SO_2 and NO_x), it also avoids significant amounts of external costs of conventional fossil fuel-based electricity generation. More and more use of wind energy should be made in order to prevent the problem of global warming.

Wind energy plants are considered a green power technology because it has only minor impacts on the environment. Wind energy plants produce no air pollutants or greenhouse gases. Wind energy is an ideal renewable energy because:

- It is a pollution-free, infinitely sustainable form of energy.

- It doesn't require fuel.
- It doesn't create greenhouse gases.
- It doesn't produce toxic or radioactive waste.

Disadvantages

Any means of energy production impacts the environment in some way, and wind energy is no different. Like every other energy technology, wind power plants do have some effects on the environment. Wind turbines cause virtually no emissions during their operation and very little during their manufacture, installation, maintenance and removal. Compared to the environmental impact of traditional energy sources, the environmental impact of wind power is relatively minor.

Wind farms are often built on land that has already been impacted by land clearing. The vegetation clearing and ground disturbance required for wind farms is minimal compared with coal mines and coal-fired power stations. If wind farms are decommissioned, the landscape can be returned to its previous condition.

The major challenge to using wind as a source of power is that the wind is intermittent and it does not always blow when electricity is needed. Wind energy cannot be stored (unless batteries are used); and not all winds can be harnessed to meet the timing of electricity demands.

Good wind sites are often located in remote locations, far from cities where the electricity is needed. Wind resource development may compete with other uses for the land and those alternative uses may be more highly valued than electricity generation. Although wind power plants have relatively little impact on the environment compared to other conventional power plants, there is some concern over the noise produced by the rotor blades, aesthetic (visual) impacts, and sometimes birds have been killed by flying into the rotors. Most of these problems have been resolved or greatly reduced through technological development or by properly sitting wind plants.

To the extent that we understand how, when, and where wind-energy development most adversely affects organisms and their habitat, it will be possible to mitigate future impacts through careful sitting decisions.

Ecological Impacts

There are two major ways that wind-energy development may influence ecosystem structure and functioning— through direct impacts on individual organisms and through impacts on habitat structure and functioning. Environmental influences of wind-energy facilities can propagate across a wide range of spatial scales, from the location of a single turbine to landscapes, regions, and the planet, and a range of temporal scales from short-term noise to long-term influences on habitat structure and influences on presence of species. The ecological influences of wind-energy facilities are complex, and can vary with spatial and temporal scale, location, season, weather, ecosystem type, species, and other factors. Moreover, many of the influences are likely cumulative and ecological influences can interact in complex ways at wind energy facilities and at other sites associated with changed land-use practices and other anthropogenic disturbances. Wind turbines cause fatalities of birds and bats through collision, most likely with the turbine blades. Species differ in their vulnerability to collision, in the likelihood that fatalities will have large scale cumulative impacts on biotic communities, and in the extent to which their fatalities are discovered. The data are inadequate to assess relative risk to passerines and other small birds. It is possible that as turbines become larger and reach higher, the risk to the more abundant bats and nocturnally migrating passerines at these altitudes will increase. Determining the effect of turbine size on avian risk will require more data from direct comparison of fatalities from a range of turbine types. The construction and maintenance of wind-energy facilities also alter ecosystem structure through vegetation clearing, soil disruption and potential for erosion, and noise. Alteration of vegetation, including forest clearing, represents perhaps the most significant potential change through fragmentation and loss of habitat for some species. Changes in forest structure and the creation of openings alter microclimate and increase the amount of forest edge. Plants and animals throughout an ecosystem respond differently to these changes. There might also be important interactions between habitat alteration and the risk of fatalities, such as bat foraging behavior near turbines.

Standardized studies should be conducted before sitting and construction and after construction of wind-energy facilities to evaluate the potential and realized ecological impacts of wind development. Pre-sitting studies should evaluate the potential for impacts to occur and the possible cumulative impacts in the context of other sites being developed or proposed. Likely impacts could be evaluated relative to other potentially developable sites or from an absolute perspective. In addition, the studies should evaluate a selected site to determine whether alternative facility designs would reduce potential environmental impacts. Post-construction studies should focus on evaluating impacts, actual versus predicted risk, causal mechanisms of impact, and potential mitigation measures to reduce risk and reclamation of disturbed sites.

Impacts on Humans

The human impacts include aesthetic impacts; impacts on cultural resources, such as historic, sacred, archeological, and recreation sites; impacts on human health and well-being, specifically from noise and from shadow flicker; economic and fiscal impacts; and the potential for electromagnetic interference with television and radio broadcasting, cellular phones, and radar. This is not an exhaustive list of all possible human impacts from wind-energy projects.

Cultural Impacts

Wind-energy facilities create both positive and negative recreational impacts. On the positive side, many windenergy projects are listed as tourist sights: some offer tours or provide information areas about the facility and wind energy in general; and several are considering incorporating visitor centers. There are two types of potential negative impacts on recreational opportunities: direct and indirect. Direct impacts can result when existing recreational activities are either precluded or require rerouting around a wind-energy facility. Indirect impacts include aesthetic impacts (addressed above) that may affect the recreational experience. These impacts can occur when scenic or natural values are critical to the recreational experience. In analyzing impacts on historic, sacred, and archeological sites, the primary concern is that no permanent harm should be done that would affect the integrity of the site. Whether or not a wind-energy project would damage the resource may depend on the specific nature of the historic resources involved.

Unlike housing developments, wind-energy projects cannot be screened from view, except behind intervening topography and vegetation. Such issues are likely to arise as wind projects are proposed in cultural landscapes, and guidance as to what constitutes an undue impact to historic or sacred sites and areas will be necessary.

Impacts on Human Health

Wind-energy projects can have positive as well as negative impacts on human health and wellbeing. The positive impacts accrue mainly through improvements in air quality, as discussed previously in this report. These positive impacts (i.e., benefits) to health and well-being are diffuse; they are experienced by people living in areas where conventional methods of electricity generation are used less because wind energy can be substituted in the regional market. In contrast, to the extent that wind-energy projects create negative impacts on human health and wellbeing, the impacts are experienced mainly by people living near wind turbines that are affected by noise and shadow flicker.

Local Economic Impacts

Wind-energy projects can have a range of economic and fiscal impacts, both positive and negative. Some of those impacts are experienced at the national or regional level; these involve, and for example, tax credits and other monetary incentives to encourage wind-energy production, as well as effects of wind energy on regional energy pricing.

Visual Impact

Landscape perceptions and visual impacts are key environmental issues in determining wind farm applications related to wind energy development as landscape and visual impacts are by nature subjective and changing over time and location.

The characteristics of wind developments may cause landscape and visual effects. These characteristics include the turbines (size, height, number, material and color), access and site tracks, substation buildings, compounds, grid connection, anemometer masts, and transmission lines. Another characteristic of wind farms is that they are not permanent, so the area where the wind farm has been located can return to its original condition after the decommissioning phase. While visual impact is very specific to the site at a particular wind farm, several characteristics in the design and sitting of wind farms have been identified to minimize their potential visual impact.

Noise Impact

Noise from wind developments has been one of the most studied environmental impacts of this technology. Noise, compared to landscape and visual impacts, can be measured and predicted fairly easily.

As with any machine involving moving parts, wind turbines generate noise during operation. Noise from wind turbines arises mainly from two sources: 1) mechanical noise caused by the gearbox and generator; and 2) aerodynamic noise caused by interaction of the turbine blades with the wind.

Experience acquired in developing wind farms suggests that noise from wind turbines is generally very low. The comparison between the number of noise complaints about wind farms and about other types of noise indicates that wind farm noise is a small-scale problem in absolute terms. Information from the US also suggests that complaints about noise from wind projects are rare and can usually be satisfactorily resolved.

Impact on Land Use

National authorities consider the development of wind farms in their planning policies for wind energy projects.

Decisions on sitting should be made with consideration to other land users. Regional and local land-use planners must decide whether a project is compatible with existing and planned adjacent uses, whether it will modify negatively the overall character of the surrounding area, whether it will disrupt established communities, and whether it will be integrated into the existing landscape. Land use planning rules in some countries recommend avoiding areas with protected designations; in others, specific areas have been earmarked for potential wind farm development.

Reduce the Negative Environmental Impacts of Wind Energy

The negative environmental impacts from wind energy installations are much lower in intensity than those produced by conventional energies, but they still have to assessed and mitigated when necessary. There are specific conditions that must be in place before an area can be considered suitable for a wind farm development. These conditions include factors such as: wind climate, topographical, logistical and ecological constraints.

A strategic environmental assessment (SEA) is the procedure used to evaluate the adverse impacts of any plans and programs on the environment. National, regional and local governments must undertake SEAs of all wind energy plans and programs that have the potential for significant environmental effects. The ecological influences of wind-energy facilities are complex, and can vary with spatial and temporal scale, location, season, weather, ecosystem type, species, and other factors. Moreover, many of the influences are likely cumulative and ecological influences can interact in complex ways at wind energy facilities and at other sites associated with changed land-use practices and other anthropogenic disturbances. Because of this complexity, evaluating ecological influences of wind-energy development is challenging and relies on understanding factors that are inadequately studied. Despite this, several patterns are beginning to emerge from the information currently available. Increased research using rigorous scientific methods will be critical to filling existing information gaps and improving reliability of predictions.

In conclusion, we must decide that if we have to produce electricity, it is certainly preferable to produce it in a way which has the smallest possible impact on the environment. From a technical and economic standpoint, the most mature form of renewable and "clean" energy is wind energy. It can effectively contribute to combating climate change while at the same time providing various environmental, social and economic benefits. On the other hand, it is necessary to minimize the impact of the wind energy, particularly in terms of environment (preservation of protected areas) and human health (noise and visual impact).

Impact of Wind Turbines on Health

Nearly 58,000 wind turbines crank out power throughout the country. These massive windmills — up to 80 feet (24 meters) tall — capture the energy in wind and convert it into free-flowing electrons that people can use to run dishwashers, air conditioning and lights.

That 8 percent may not sound like much until you realize that wind power is just catching on in the United States. Huge new wind farms accounting for thousands more megawatts of capacity are in development as we speak, and the U.S. Department of Energy set goals in its 2017 Wind Vision: A New Era for Wind Power in the United States. Its aim is to have wind provide, 20 percent by 2030 and 35 percent in 2050. Wind power already covers 14 percent of the electricity demands in the European Union.

Until recently, there were three main issues regarding the possible downsides of wind power: bird and bat deaths, cost, and disrupting the appearance of natural landscapes.

But a new objection to wind power has popped up in again in the news – the idea that wind power endangers the health of people who live near windmills. Some people call this theory "wind-turbine syndrome".

Those concerned about wind-power syndrome are interested in finding out if and how wind power could be making people sick. Is everyone living near windmills facing health problems? Let's take a look at the possible health risks associated with wind farms and find out whether we should be worried about the steady increase in wind-generated power throughout the world.

Many residents living near wind farms like this one in Australia complain about noise and other health problems.

The rapidly spinning blades of huge wind turbines have an effect on their surroundings, and it goes beyond aesthetics. The rapidly spinning blades can produce a weak but distinctive noise, as well as disruptions in air pressure. The noise is generated by the movement of the blades through the air, as well as from the from the turbine machinery.

Infrasound, is sound that is lower in frequency than 20 Hz or cycles per second. That's the "normal" limit of human hearing. infrasound is the primary issue for those concerned about wind-turbine syndrome. They also say that audible sound and vibrations coming from wind turbines contribute to the health problems reported by some people who live close to wind farms. Symptoms of wind-turbine syndrome might include:

- Headaches.
- Sleep problems.
- Night terrors.
- Ringing in the ears (tinnitus).
- Mood problems (irritability, anxiety).
- Concentration and memory problems.
- Issues with equilibrium, dizziness and nausea.

These symptoms have been observed and documented by a limited number of scientists studying small groups of people, and the scientific community hasn't concluded whether wind-turbine syndrome exists. There are also mixed opinions on whether wind turbines emit infrasound and if the amount is any more than that emitted by diesel engines or waves crashing on the beach. But we do know that at high speeds, wind turbines can produce an audible hum and vibration that can be carried through the air.

Wind Turbine Syndrome and Vibroacoustic Disease

Industrial wind turbine development is being strongly supported by private interests due largely to government subsidies and artificial financial incentives, making wind turbine development possible, whereas otherwise it would fail on financial grounds alone. There are currently no government regulations affecting placement of industrial wind turbines (setback) in Massachusetts. However, many concerns have been expressed by medical doctors around the world regarding the harmful effects to humans of operating industrial wind turbines.

Recently published peer reviewed comprehensive information prepared by medical doctors, pathologists and engineers strongly warn of major physiological consequences of living too close to industrial wind turbines. These have been classified into two categories according to two different phenomena associated with wind turbine noise. One is Wind Turbine Syndrome (WTS), which produces several symptoms related to the vestibular system's (balance) organs: disturbed sleep, headaches, tinnitus (ear ringing), and sense of quivering or vibration, nervousness, rapid heartbeat, nausea, difficulty with concentration, memory loss, irritability and anger. The other is Vibroacoustic Disease (VAD), which causes direct tissue or organ damage. WTS symptoms discontinue when the person moves away from the source; however, VAD symptoms continue long after the source of infrasound is turned off. Both WTS and VAD can be very harmful and debilitating and possibly deadly.

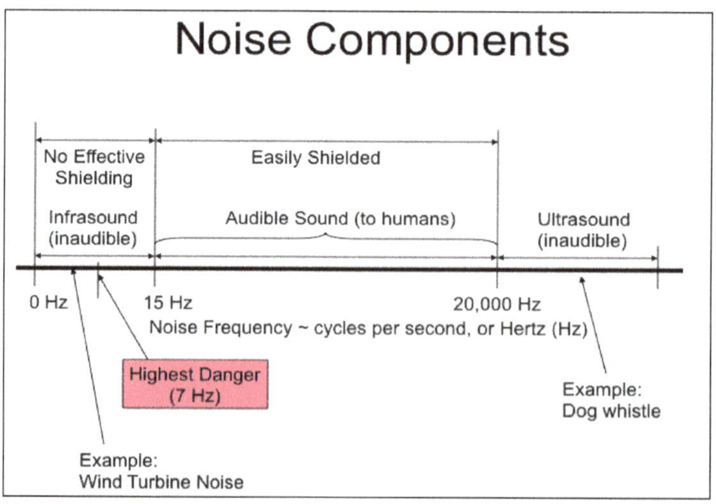

Components of noise, from low frequency to ultrasound.

Wind turbine noise is comprised of two different frequency sets. The so-called whooshing noise (which is audible) is at approximately 300 Hz. Infrasound (which is inaudible) occurs at frequencies from 0 to 15 Hz. WTS is caused by the former and the latter; VAD is caused by the latter.

WTS is a term coined by Dr. Nina Pierpont. Her book, is an excellent resource regarding health effects wind turbine generated WTS on the vestibular system. It provides an excellent summary in medical terms and laymen's language of the dangers of WTS and no attempt is made to expand on her presentation here.

VAD, however, is less well known to the public, but has been proven that man-made machines can cause VAD and can kill,. The main question is whether there is sufficient amplitude (energy) in wind turbine infrasound to cause VAD. The following supports this possibility:

"These results comparison of infrasound levels contained in a residence in close proximity to wind turbines to a known VAD case caused by a nearby port grain terminal – the former being much larger irrefutably demonstrate that wind turbines in the proximity of residential areas produce acoustical environments that can lead to the development of VAD in nearby home-dwellers".

"Another source of ILFN has appeared: wind turbines ILFN levels contaminating the home of Case situated near wind turbines are amply sufficient to cause VAD widespread statements claiming no harm is caused by in-home ILFN produced by wind turbines are fallacies that cannot, in good conscience, continue to be perpetuated. In-home ILFN generated by wind turbines can lead to severe health problems, specifically, VAD".

All things in life have natural frequencies, e.g., buildings, bridges, guitar strings, our internal organs. If the frequency of a forcing function, e.g., noise, happens to match the natural frequency of an object, the object will react at that same frequency and motion of the object will be amplified. Most people recall from high school the films of the infamous Tacoma Narrows Bridge ("Galloping Gertie") which was built in 1940, and collapsed four months later due to large harmonic (synchronous) oscillations caused by wind. VAD can cause similar catastrophes within our bodies. The resonant frequencies of our body organs happen to be low. Industrial wind turbines happen to produce low frequency noise (1-2 Hertz – the blade passage frequency) with overtones of up to about 20 Hz. This noise is transmitted through every medium between the turbine and your body. It is well documented that infrasound generated by certain machines can cause VAD.

Researchers also firmly states that "real and efficient zoning for wind turbines must be scientifically determined, and quickly adopted, in order to competently and responsibly protect Public Health." Potential dangers to the health and safety of the public cannot be ignored, and clearly there are well researched and published papers that conclude that wind turbines can cause WTS and VAD. Infrasound has the nasty characteristic of traveling great distances and shielding of infrasound is ineffective. Until we have a determination of required setbacks to avoid health effects from wind turbine generated infrasound, erection of wind turbines should be halted. The establishment of such setbacks should not be a burdensome request:

"Existing evidence is not sufficient to make several important quantifications, including what portion of the population is susceptible to the health effects from particular exposures, how much total health impact wind turbines have, and the magnitude of exposure needed to cause substantial risk of important health effects. However, these are questions which could be answered if some resources were devoted to finding the answer. It is not necessary to proceed with siting so that more data can accumulate, since there is enough data now if it were gathered and analyzed,".

The State Department of Public Health of the Commonwealth of Massachusetts is responsible for ensuring public health and safety. Development of wind turbine technology has progressed more rapidly than our knowledge of the potential consequences to human health and safety. The Commonwealth needs to place a moratorium on this development until conservative setback limits (2-3 miles) consistent with medical findings are established for proximity to homes, industries, schools, hospitals, etc. For proximity to wild animals, no practical setback can be established.

Impact on Water Resources

A wind energy project can impact surface water and groundwater in several different ways, including the use of water resources, changes in water quality, alteration of the natural flow system, and the alteration of interactions between the groundwater and surface water. For the most part, however, wind energy development does not require much water, except during the construction phase and, to a lesser extent, during decommissioning. These water uses are temporary, and during the operations phase, water use would be minimal.

Site Monitoring and Testing

Generally, the impacts during site monitoring and testing would be relatively limited because new access roads might not be needed, on-site activities would be limited and temporary, and the size of the work crew would be small. As a result, very little, if any, water would likely be used during this phase of development. If water was needed, it would probably be trucked in from off site. Impacts to water resources, local water quality, water flows, and surface water/groundwater interactions are expected to be negligible to small, unless extensive excavation or road construction occurs.

Site Construction

Most of the impacts on water resources would occur during construction.

Use of Water Resources

A number of construction activities would use water. Because the construction phase may last more than 1 year, potentially large amounts of water would be needed. The water may be trucked in from off site or obtained from local groundwater wells or surface water bodies near the facility, depending upon the availability of those sources. Activities related to use of water resources would include:

- Water used for dust control during the construction of access roads, clearing of vegetation, grading, and road traffic;
- Water used for making concrete used in the foundations of wind towers, substations, central control buildings, and various personnel support facilities; and
- Water used by the construction crew.

Water Quality

Many construction activities associated with a wind energy development project could alter the quality of surface water and, to a lesser extent, groundwater. These include:

- Activities that aggravate soil erosion, such as activities that disturb the ground surface, heavy equipment traffic, activities that alter surface runoff patterns, and extraction of geologic materials from borrow areas or quarries;
- Weathering of freshly exposed soil or spoils from foundation excavation, quarry or borrow pit operations, or access road construction, which would release various chemicals through oxidation and leaching processes;
- Discharges of wastewater or sanitary water; and
- Pesticide application, unless use is limited to nonpersistent, immobile pesticides and applied only in accordance with label and application permit directions and stipulations.

Alteration of Water Flow Systems

Natural surface water and groundwater flow systems could be potentially impacted by construction activities. Surface water flows may be diverted on site and off site by access road systems or storm water control systems. Excavation activities or geologic material extraction may alter surface overflow and groundwater flow. The withdrawal of surface water and groundwater for water uses and the discharge of wastewater and storm water would also affect the water flows of the surface water bodies and groundwater.

Alteration of Surface Water/Groundwater Interaction

Construction activities could alter the interaction between surface water bodies and local groundwater in systems where the two resources are hydrologically connected. In these circumstances, extracting water from one source eventually could affect the other source as well. Similarly, altering the water quality of one source could affect the water quality of other sources at downgradient locations. Impacts also could occur if construction activities (e.g., excavation, blasting, trenching) create a conduit between a surface water body and a groundwater aquifer, or between two aquifers, by breaching the hydrologic barrier. This could result in unwanted dewatering or recharge of any of these water resources, depending on local hydrogeologic conditions. In addition, storm water control systems and any other activity that alters the ground surface could affect groundwater infiltration as well as the response time of a nearby surface water body.

Site Operation

As various construction and related activities diminish, the environment will re-establish a new equilibrium. If appropriate mitigation measures are implemented during the construction phase, potential impacts to water during site operation would be limited to the degradation of water quality as a result of improper pesticide use or vehicle traffic.

Site Decommissioning

The impacts on water resources during decommissioning would depend on the decommissioning activities involved. Such activities may involve removal of all access roads, on-site roads, transformer pads, and building foundations. Originally disturbed land areas would likely be restored to their original grade and revegetated. Water wells may be abandoned in place. The potential impacts would largely be the same as those described for the construction phase.

Mitigation Measures

Potential water resource impacts would mostly occur during the site construction and decommissioning phases. Mitigations measures that could reduce such impacts include:

- The size of cleared and disturbed lands should be minimized as much as possible. Existing roads and borrow pits should be used as much as possible.
- Topsoil removed during construction should be salvaged and reapplied during reclamation. Disturbed soils should be reclaimed as quickly as possible or protective covers should be applied.
- Operators should identify unstable slopes and local factors that can induce slope instability (such as groundwater conditions, precipitation, earthquake

activities, slope angles, and dip angles of geologic strata). Operators also should avoid creating excessive slopes during excavation and blasting operations. Special construction techniques should be used where applicable in areas of steep slopes, erodible soil, and stream channel/wash crossings.

- Erosion controls that comply with county, state, and federal standards should be applied. Practices such as jute netting, silt fences, and check dams should be applied near disturbed areas.
- Operators should gain a clear understanding of the local hydrogeology. Areas of groundwater discharge and recharge and their potential relationships with surface water bodies should be identified.
- Operators should avoid creating hydrologic conduits between two aquifers during foundation excavation and other activities.
- Proposed construction near aquifer recharge areas should be closely monitored to reduce the potential for contamination of said aquifer. This may require a study to determine localized aquifer recharge areas.
- Foundations and trenches should be backfilled with originally excavated material as much as possible. Excess excavated material should be disposed of only in approved areas.
- Existing drainage systems should not be altered, especially in sensitive areas such as erodible soils or steep slopes. When constructing stream or wash crossings, culverts or water conveyances for temporary and permanent roads should be designed to comply with county standards, or if there are no county standards, to accommodate the runoff of a 10-year storm. Potential soil erosion should be controlled at culvert outlets with appropriate structures. Catch basins, roadway ditches, and culverts should be cleaned and maintained regularly.
- On-site surface runoff control features should be designed to minimize the potential for increased localized soil erosion. Drainage ditches should be constructed where necessary but held to a minimum. Potential soil erosion should be controlled at culvert outlets with appropriate structures. Catch basins, drainage ditches, and culverts should be cleaned and maintained regularly.
- Pesticide use should be limited to nonpersistent, immobile pesticides and should only be applied in accordance with label and application permit directions and stipulations for terrestrial and aquatic applications.

Permissions

All chapters in this book are published with permission under the Creative Commons Attribution Share Alike License or equivalent. Every chapter published in this book has been scrutinized by our experts. Their significance has been extensively debated. The topics covered herein carry significant information for a comprehensive understanding. They may even be implemented as practical applications or may be referred to as a beginning point for further studies.

We would like to thank the editorial team for lending their expertise to make the book truly unique. They have played a crucial role in the development of this book. Without their invaluable contributions this book wouldn't have been possible. They have made vital efforts to compile up to date information on the varied aspects of this subject to make this book a valuable addition to the collection of many professionals and students.

This book was conceptualized with the vision of imparting up-to-date and integrated information in this field. To ensure the same, a matchless editorial board was set up. Every individual on the board went through rigorous rounds of assessment to prove their worth. After which they invested a large part of their time researching and compiling the most relevant data for our readers.

The editorial board has been involved in producing this book since its inception. They have spent rigorous hours researching and exploring the diverse topics which have resulted in the successful publishing of this book. They have passed on their knowledge of decades through this book. To expedite this challenging task, the publisher supported the team at every step. A small team of assistant editors was also appointed to further simplify the editing procedure and attain best results for the readers.

Apart from the editorial board, the designing team has also invested a significant amount of their time in understanding the subject and creating the most relevant covers. They scrutinized every image to scout for the most suitable representation of the subject and create an appropriate cover for the book.

The publishing team has been an ardent support to the editorial, designing and production team. Their endless efforts to recruit the best for this project, has resulted in the accomplishment of this book. They are a veteran in the field of academics and their pool of knowledge is as vast as their experience in printing. Their expertise and guidance has proved useful at every step. Their uncompromising quality standards have made this book an exceptional effort. Their encouragement from time to time has been an inspiration for everyone.

The publisher and the editorial board hope that this book will prove to be a valuable piece of knowledge for students, practitioners and scholars across the globe.

Index

CPSIA information can be obtained
at www.ICGtesting.com
Printed in the USA
LVHW060908030122
707669LV00002B/411